Partisan Histories

Partisan Histories

The Past in Contemporary Global Politics

Max Paul Friedman and Padraic Kenney

PARTISAN HISTORIES

First published in 2005 by
PALGRAVE MACMILLAN™
175 Fifth Avenue, New York, N.Y. 10010 and
Houndmills, Basingstoke, Hampshire, England RG21 6XS
Companies and representatives throughout the world.

PALGRAVE MACMILLAN is the global academic imprint of the Palgrave Macmillan division of St. Martin's Press, LLC and of Palgrave Macmillan Ltd. Macmillan® is a registered trademark in the United States, United Kingdom and other countries. Palgrave is a registered trademark in the European Union and other countries.

ISBN 1–4039–6455–6

Library of Congress Cataloging-in-Publication Data

Friedman, Max Paul.
    Partisan histories : the past in contemporary global politics / Max Paul Friedman and Padraic Kenney.
      p. cm.
    Includes bibliographical references and index.
    ISBN 1–4039–6455–6—ISBN 1–4039–6456–4 (pbk.)
      1. Public history. 2. Memory—Social aspects. 3. Nationalism—History—20th century. 4. History—Philosophy. 5. World politics—20th century. I. Kenney, Padraic, 1963– II. Title.

D16.163.F75 2005
900—dc22   2004061671

A catalogue record for this book is available from the British Library.

Design by Newgen Imaging Systems (P) Ltd., Chennai, India.

First edition: June 2005

10 9 8 7 6 5 4 3 2 1

Printed in the United States of America.

Contents

Introduction: History in Politics

Max Paul Friedman and Padraic Kenney

The novelist William Faulkner once observed, "The past is not dead; it is not even past." Events that took place years or centuries ago directly affect the present by setting the conditions in which today's events unfold. Moreover, the way we tell stories about the past can sway our thinking about the present. Consider two interpretations of the same event: "The Vietnam War was a disastrous mistake, never to be repeated"; "The Vietnam War was a noble cause; next time we must have the will to win." These two versions of the same past represent competing histories and have very different implications for the present. Thus history, the meaning we assign to the past, can influence such momentous decisions as whether or not to go to war.

For any given subject there is no single, true version of history but multiple contending ones. Often these histories are produced with an immediate goal in mind: they are *partisan histories*, narratives about the past designed to help win arguments and political struggles.

This book examines the role of competing interpretations of the past in political conflicts in different parts of the world today. Each of the nine cases focuses on arguments about history in democratic countries. We have selected democracies because authoritarian regimes, generally, do not witness debates over history, seeking instead to impose their own official versions of history upon a population that may variously accept them, resist, or privately nurture alternative versions to the state or party line. Democracies differ in at least two crucial ways: they provide public space for lively debates, and rather than rely on coercion to obtain cooperation from their populations, they are governed by groups that are able to present a compelling case for their legitimacy as rulers. Often, the evidence offered to support these claims is rooted in a vision of history. Thus democracies not only permit, but also foster or even require, disputes over history.

This is not a book about the work of historians, however. For at least two reasons, the disputes are not between a scientific, objective, professional history produced by historians, on the one hand, and a mythic, politicized, or invented history proclaimed by politicians, on the other. The first reason is that historians are not in agreement with one another. The second is that they, too, carry out their work under the influence of their own values and interests and those of the societies in which they live. Although good historians abide by professional standards of evidence and documentation and strive for objectivity—that elusive but "noble dream"[1]—they would be mistaken to believe that they ever achieve it. As Ronald Grigor Suny writes in his contribution to this volume, the "pretension to objectivity is also a pretension to an untroubled authenticity of a single reading," to the exclusion of alternative interpretations. Objectivity is a laudable goal that can never be reached. Claims to complete objectivity often veil a desire to silence competitors.

Rather than being scholarly disputes among specialists, the partisan histories described here are fought out in a public arena outside the groves of academe for political goals of great consequence. Indeed, when the contributors to this volume gathered around a conference table in New York in October 2003, the conversation turned to their experiences of jail, death threats, and exile—experiences familiar to historians under some regimes. It may seem surprising that history can be a dangerous business. As the essays that follow will show, the stakes in history can be very high: war or peace, freedom or imprisonment, powerlessness or control of resources or even of the national government. The actors are a diverse crowd, including politicians, diplomats, ethnic cleansers, and war criminals; revolutionaries, university students, intellectuals, moviemakers, and even comic book writers. All recognize that the past is a rich source of stories, images, metaphors, and "lessons" that have compelling power over the imagination and can move people to action. In the national contests for power depicted here, history becomes a weapon in the struggle for symbolic capital, wielded to acquire legitimacy for one's own side while delegitimizing the opposition.

To think clearly about these ideas, we need to understand what we mean by several key terms. The *past* is the closest we have to an objective concept: it is simply "what happened before," an infinite number of events that would be impossible to catalog and reproduce with accuracy because to do so would require omniscience, omnipresence, and eternity. *History* is a sustained narrative about the past, a narrative that is neither natural nor scientific, but is carefully constructed to give meaning to past events by selecting some for inclusion, leaving others out, and interpreting the ones that are recounted in order to convey certain conclusions. To say that history is subjective is not to say that it is illegitimate, or not useful, or unscholarly; it is merely to recognize the production of history as a human activity. Although all history is in some sense political, the *partisan histories* analyzed in these essays are narratives that play central roles in national or international conflicts, without which those conflicts cannot be understood. In their essay in this volume, Subho Basu and Suranjan Das invoke E.J. Hobsbawm's distinction between subjective partisanship in historical writing and partisan history. Whereas the former "rests on disagreement not about verified facts, but about their selection and combination, and about what may be inferred from them,"[2] in the latter, advancing political interests is more important than standards of evidence. Although we argue that all history is subjective, this is not to suggest a kind of relativism. The very best histories adhere to high methodological standards. But even they are not free of political implications.

In the last three decades, a new subdiscipline of the humanities has emerged to examine some of the questions we engage with here: the study of *memory*.[3] One tendency is to offer the term as a counterpoint to scholarly or official histories: memory as an experientially based interpretation of the past not found in books but sustained across time through cultural practices such as rituals, memorials, and folklore. The literal meaning of the word derives from a mental faculty that is private and individual: in a concrete sense, people have memories, peoples do not. Two French scholars contributed importantly to broadening the definition of the term to apply it to society. Maurice Halbwachs pointed out in the 1920s that individual memories do not grow in isolation, but are influenced by interaction with others. His understanding of collective memory remained rooted in the organic mental process: memory could not last more than a lifetime, but it could be shared by a community of individuals with common experiences.[4]

In the 1970s, Pierre Nora drew on Halbwachs's concept of collective memory to describe something more enduring and even "living," a presence of the past borne not only by the recollections of a group, but by physical sites and cultural practices that last much longer than a lifetime. Unlike the allegedly dry, lifeless, elitist history written by historians, memory in this sense is supposed to be part of the living body of society, and it is essential to the constitution of national identity.[5] This interpretation, though inspired by a democratic impulse to recapture the study of the past from elitist scholars, paradoxically lends itself well to romantic nationalist myths of origin.

In this volume, we do not directly engage in academic arguments about the usefulness of the term "memory." Instead, we use the term to designate diffuse representations of the past that need not be textual. Memory does not emerge organically from some mythical living body of a nation, but is shaped by many forces including lived experience, and also by the deliberately articulated versions of the past we call history. There is no bright shining line between history and memory in the essays that follow, but histories are usually consciously presented in some form, whereas collective memory can be a more passive understanding of the past. However, we do not believe memory to be an apolitical, organic process, any more than we believe that nations are natural entities. As Patrick Geary has written, all memory is "memory *for* something."[6]

Political in its basic sense means, "related to power." We recognize that power can be found in many sites: in language, in symbols, in personal relationships, in everyday practices. In this volume, the authors focus especially on a more narrow and public realm of the political: the contest for control or influence over the state and its resources. Readers of comparative political science, accustomed to studies of politics at this level, often find discussions of history confined to an introductory "background" section. But history is not mere background, something one absorbs before coming to grips with more current concerns: it is often the very stuff of contemporary political conflict, as all our cases demonstrate.

We have deliberately selected asymmetric cases from five continents to compare with one another. Comparison is a useful tool of analysis whose purpose here is not to conflate the different examples into an overarching global phenomenon, but to highlight differences and commonalities. One might not expect to find much similarity among contemporary political conflicts drawn from Europe, Asia, Africa, and the Americas, but comparing such divergent areas and their unique experiences does yield insight into the way the past can be instrumentalized for political purposes in a broad range of circumstances.

The most obvious conclusion to be drawn from this comparative look is that whether one examines nation-building projects in Nigeria, electoral contests in Spain, foreign policy in the United States, or the Arab–Israeli peace process, accounts of the past are inextricably linked to the politics of the present. For those interested in assessing rival histories, an awareness of their positioning in political conflicts is essential. For those interested in understanding contemporary political conflicts in various countries, recognizing the role played by partisan histories is indispensable to cutting through mythmaking and pretensions of absolute, objective truth.

It is not surprising that in contests for control over the resources of the nation-state, groups should seek legitimacy and adherents by presenting themselves as the only true representatives of the nation through historical narratives that support that claim: the rationale for nationalism is always sought in history. This is sharply evident in the case of so-called new nations such as Armenia or Nigeria whose boundaries were established recently enough to require justification or whose populations are diverse enough to require a unifying narrative explaining why the people owe allegiance to the state. But competing nationalist discourses are also explicitly or implicitly present in long-established nations in contemporary conflicts over issues not directly connected to borders or membership in the national community. Germany's postwar effort to regain national legitimacy took the form of various attempts at restitution, expiation, and distancing from the Nazi era, a process marked by the incessant clash of partisan histories wielded by competing sectors of German society. Chile and Spain face the dilemmas not of new nations but of new regimes, where democratic rule has replaced dictatorship in a peaceful transition made possible initially by agreements not to aggressively prosecute or even officially commemorate the crimes of the past, the "pacts of silence." However, as the moment of transition has receded and the pacts have eroded, political parties from different parts of the spectrum have presented partisan histories that make claims to different elements of the national narrative to advance their cause.

One pattern that emerges from the cases is a disparity between left-wing and right-wing approaches to national self-examination. Nationalist conservatives tend to avoid or oppose searching explorations of national guilt or responsibility because they prefer to put forth a celebratory narrative of tradition, of legitimacy through continuity and links to ancestors; if those ancestors were criminals, the claim to legitimacy is weakened. If nationalist conservatives do not wish to take ownership of those crimes, they have to deny that they happened (as when right-wing Germans refuse to believe that regular army units were capable of committing atrocities), denigrate their importance (as when right-wing Japanese belittle the suffering of sex slaves), or attribute them to extenuating circumstances such as military exigencies and self-defense against the victims (conservative American dismissal of civilian deaths in Vietnam, Chilean claims that General Augusto Pinochet's victims were Marxist rebels). Left-liberals tend to be more skeptical of such claims of descendance and often celebrate change rather than tradition, so they have less interest in presenting an unblemished national past. Moreover, by insisting on public attention to atrocities associated with conservative predecessors, they discredit the celebratory nationalism used by contemporary conservatives, and by embracing national responsibility for the past, they can appeal to sectors of the population that identify with the victims while at the same time furthering a cosmopolitan agenda.[7] This is not to suggest that the Left has no other goals in its uses of the past. Andrew Beattie's essay notes the sometimes scattershot charges of West Germany's links to its fascist past made by the Left, and analyzes East Germany's instrumentalization of memories of communist resistance to the Third Reich. Ilan Pappe implicates the Israeli "Peace Camp," generally associated with the Labor Party, in obscuring the events of the 1948 partition of Palestine. Carsten Humlebæk reports that in Spain, the Socialist Party was the first to break a truce over the use of history for partisan purposes when it feared losing a national election in 1993. Nevertheless, claims of an untarnished national history are almost by definition of greater concern to those for whom asserting national pride based on continuity with the past constitutes a central part of their political program.

In these disputes, the role of intellectuals varies and is rarely so marginal as many self-deprecatingly believe, nor so impressive as some might wish. In countries such as Israel and Germany where professional historians have substantial access to the national media and take part in national political debates, their interpretations of recent history can affect political developments directly. In the United States, Japan, Armenia, and Nigeria, the work of professional historians seems to have less impact on national discussions than do works of fiction, film, public memorial sites, and claims by politicians. Popular writers outside the academy and historians who sustain rather than critique triumphalist nationalist myths tend to have the most success with the reading public and the national media. One reason this is of interest (not only to academics wistful for a larger audience) is that much academic discourse nowadays acknowledges the artifice of nationalism and is skeptical of heroic myths. Many academics and liberals take it for granted that nationalism has been a damaging force, and they cite abundant examples of violent consequences, from Hitler's Germany to Milosevic's Serbia, or the clashes between Israelis and Palestinians, Indians and Pakistanis. But as Toyin Falola points out in his essay on Nigeria, nationalism can also be a positive force in helping to draw a disparate population together.

Finally, comparing these disparate cases shows that although every case is unique, we can detect phenomena that connect them and speak to a kind of *Zeitgeist* (spirit of the age) based on transnational developments. Germany's confrontation with its Nazi past appears as a positive or negative model in nearly every country coming to terms with its own national traumas. As John Torpey has written, "one might well say that 'we are all Germans now' in the sense that all countries . . . that wish to be regarded as legitimate confront pressures to make amends for the more sordid aspects of their past."[8] The Holocaust and its aftermath loom large in many parts of the globe; for example, the reports issued by truth commissions examining atrocities committed in the "dirty wars" in Argentina, Brazil, Uruguay, and Guatemala were all titled *Nunca más*, "Never again." The Nuremberg trials of German war criminals conducted by the Allies after World War II (and, to a less visible extent, the Tokyo trials of Japanese war criminals) were central in establishing concepts of international law and human rights that influence contemporary investigations. Nuremberg enshrined the notion of "crimes against humanity," crimes so heinous that they should be subject to no statute of limitations and can be prosecuted even if the acts were officially regarded as legal or condoned by the previous regime. The tribunal also established the principle of "command responsibility," that is, that superiors may be judged guilty even if they did not participate directly in the crime, while rejecting the excuse "I was just following orders" as a defense for criminal acts. At the same time, Nuremberg and Tokyo are sometimes held up as a negative model of "victor's justice" by people eager to ensure that today's proceedings against former regime supporters are carried out as part of an internal process of national renewal, not imposed by conquerors. More recently, South Africa's post-Apartheid Truth and Reconciliation Commission has served as a model for similar commissions in other countries.

These transnational developments are in part due to the demonstration effect of a few highly publicized cases such as Germany and South Africa, and are also influenced by external actors. German politicians debating forms of restitution for Nazi crimes monitor public opinion in the United States and seek to maintain harmonious relations with Israel and neighboring countries in Eastern Europe. International pressure from East Asia and beyond seems to account for the increasingly pervasive discourse of apology in Japan. In truth commissions in El Salvador and Guatemala, the presence of United Nations personnel served as a counterweight to the preponderance of power held by the military and the right wing that would have preferred to see no investigation of the dirty wars. In Spain, the Left's decision to break the pact of silence over the civil war of 1936–1939 came after a delegation to Mexico found that Spain's war was already being commemorated there.

Where partisan histories are deployed in favor of or in opposition to reparations, the core issue is rarely the monetary damages themselves, since the damage done by atrocity can never be repaired or repaid. More important is the need for official recognition of past wrongs as a form of moral compensation and a way of bringing renewal to a society whose past has delegitimized it internally and internationally. There must be some form of punishment of or contrition by the perpetrators or their heirs to satisfy the needs of the victims and their allies in civil society and the international community. To say that this is a political process is not to denigrate those who participate in it, but merely to acknowledge how decisions are reached in democracies: politically. Where the process includes some version of a truth commission, the point of these investigations is not simply to reveal what is in any case often already known. The goal is, above all, official acknowledgment as a corrective to previous official denials and deception (sometimes embodied in the physical disappearance of victims).[9] These are some of the commonalities that emerge from a study across multiple cases.

In a sense, all histories are political, but not all partisan histories are equal. Some are more plausible than others. While the historical profession provides standards of evidence and argumentation and forms of peer review to evaluate academic works of history, these, too, are implicated in political processes for good or ill. Ultimately, whether an interpretation flourishes in the public sphere is determined not by guardians of an academic discipline but by the broader political context of the society in which it appears. The chapters of this book should make this apparent.

The Cases

The nine essays that follow are divided into three parts. The first group of cases are societies riven by present conflicts over past crimes by former regimes, including

Germany, Japan, Chile, and Spain. The second group consists of "new nations" (postcolonial or newly independent states) where contested narratives of the past are an integral part of ongoing disputes over borders and sovereignty (India and Pakistan, Israel and Palestine) or national legitimacy and cohesion (Armenia and Nigeria). The third part examines the uses of historical analogies in foreign policy, especially the role such analogies played in the United States in debates over the meaning of the Vietnam War and the subsequent use of military force. Although each case is unique, and interesting for its particular circumstances, there is naturally some overlap among the categories.

Andrew Beattie's essay on Germany provides a natural starting point because the German process of *Vergangenheitsbewältigung*, "coming to terms with the past," so often serves as a reference point and a model, whether positive or negative, invoked in other countries. Germany is sometimes described as confronting a "double past," working through the aftermath of both the Nazi era and communist rule in East Germany. Merely asserting such equivalence between two very different dictatorships is highly controversial, but there is no denying the complexity of the web of historical narratives running through German political life.

Germans after World War II faced an array of challenges whose working through would require explicit engagement with the past. If self-aggrandizing German nationalism was thoroughly discredited by Nazi crimes, on what basis could Germans find legitimacy as a nation? Beattie examines the two German states' distinct and mutually exclusive legitimation strategies, each based on a disavowal of National Socialism. In the East, this took the form of a doctrine of "anti-fascism" that linked the capitalist West to the Nazi era while celebrating communism as the true source of resistance. In the West, the doctrine of "anti-totalitarianism" fused communism and fascism into a single phenomenon, with democracy its antithesis. Since many East German communists, including leaders such as Walter Ulbricht and Erich Honecker, had survived Nazi persecution, there was a plausible basis for the assertion of a historical antifascism that was carefully nurtured through official discourse, commemorations, and educational policy. West Germany's democratic bona fides were established through a series of restitution and compensation payments to Israel and survivors of Nazi crimes, a foreign policy based on international cooperation, and, especially after the 1950s, an intense and ongoing critical debate in the public sphere. Since the end of the cold war, that debate has intensified.

In postwar Japan, professional historians for many years steered clear of the recent past in favor of studying earlier periods, leaving the terms of the debate over Japanese responsibility and possible restitution to be set by political parties that generally min-imized the importance of Japanese crimes, and by popular forms of writing, includ-ing best-selling comic books, that promote a revisionist nationalism. These are not the only sources of information available in Japan, where the reading public buys serious books about the war and leftist schoolteachers in their lectures expand beyond the meager treatment of World War II contained in official textbooks. As Alexis Dudden makes clear in her essay, the impression widely held in the West of an unapologetic Japan is out of date. The country does not seem to be developing anything on the scale of an official penitential discourse like the one in Germany that sets boundaries for acceptable behavior by public figures. Yet in today's Japan, the concept of apologizing for past crimes has moved from being a faintly audible demand from the victims to being co-opted by the state in an attempt to meet contemporary international standards of legitimacy that require gestures of atonement. In both former members of the Axis, while individuals may be impelled by some degree of shame or ethical motivation to make public conciliatory gestures, material factors such as the drive for access to important export markets also influence government policies designed to improve the national image.

Katherine Hite's essay on Chile presents another post-dictatorship society, but one whose former regime was not defeated in war. A number of repressive Latin American military regimes lost power in the 1980s and 1990s, but this was a retreat "in an orderly fashion," with the departing generals able to impose conditions that prevented successor governments from prosecuting former leaders guilty of crimes.[10] This was the Faustian bargain made by post-authoritarian governments: trading amnesty for stability, they accepted a form of extortion according to which high military officers consented to live under democratic rule only on condition that they not be brought to justice, implicitly threatening to revolt if their demands were not met. Those implicated in former crimes retain a degree of power that constrains democracy and the rule of law, while they seek to influence interpretations of the past that become official history.

During Chile's transition to democracy, the Chilean military, which had not just lost a war, retained powerful allies among conservative civilians. Chile therefore saw few prosecutions, but pressure from victims and their families, and from opposition political parties, led to the creation of a truth commission that presented detailed accounts of the structure of repression under General Augusto Pinochet. Hite's study delves deeply into the intersection of official histories and private memories of what she calls the "trauma" of experiences associated with the overthrow of the democratic government of Salvador Allende in 1973 and the subsequent dictatorship under Pinochet. This trauma still circumscribes electoral politics in Chile today. An uneasy truce and sporadically broken silence endured through most of the 1990s, until Pinochet retired from the military, entered the Senate, and then was sought for prosecution by a Spanish magistrate. Loosely organized groups of young people whose parents were victims of the Pinochet regime played an important role in breaking the silence when they staged demonstrations in front of the homes of perpetrators. (This generational conflict is not unusual; many young Germans of the "generation of '68" confronted their parents over their roles in the Third Reich and produced a sea change in national discourse on the past.) As Hite concludes, since private memories endure in spite of any agreements laying out the limits of official history, political leaders may be forced to come to terms with popular demands for addressing the past.

The bitter legacy of the Spanish Civil War of 1936–1939 remained largely unprocessed in the constricted public sphere of Spanish society under the dictatorship of Francisco Franco, whose rule ended with his death in 1975. Unlike what happened in Japan and Germany, but as in Chile, there was no "de-Francoization"; the principal civil and military institutions of the Franco regime remained in place, and there were no prosecutions of former officials. There followed a period of transition in which the Spanish military and political parties respected a "pact of silence," out of fear that to begin a heated debate over the civil war might risk sparking a new one. To maintain social peace, the parties agreed to apply an amnesty program to political prisoners who had opposed the dictatorship as well as to Franco's enforcers who had committed human rights violations.

But the amnesty threatened to become a kind of amnesia, writes Carsten Humlebæk in his essay. Although all political parties were eager to protect the fragile democracy, it became clear that the benefits of the pact of silence were unequally distributed. The pact worked in favor of the Right, whose misdeeds were obscured, but not of the Left, whose suffering and resistance in the war and the dictatorship could neither be compensated for nor turned into political capital because the subject was taboo. In 1993, after enough time had passed that democratic rule seemed stable and the Socialists thought they might lose an election, they abandoned the pact and tried explicitly to associate the rival conservative party (the Partido Popular) with Franco's dictatorship. A steady stream of challenges to the pact of silence followed, from restitution claims by former prisoners or victims' families to the successful campaign to grant Spanish citizenship to aging foreign volunteers on the anti-Franco side in the civil war. This process of step-by-step rectification of past wrongs has included the search for new forms of expressing Spanish nationalism without echoing Franco's exaltation of the "Spanish Nation" or his practice of suppressing powerful regionalist sentiment. One recently proposed model reflects the influence of the German concept of constitutional patriotism, that is, civic national pride not based on ethnicity or a myth of origin, but rather on the satisfaction of sustaining a democratic form of governance.

Comparing the cases of Germany, Japan, Chile, and Spain shows that in post-authoritarian societies, the way new governments establish their legitimacy and the scope of discourse about the past is greatly influenced by the fate of the perpetrator regime, how discredited it has become, and how much political power its remnants may yet hold.

If politics works partly through history in countries coming to grips with the legacy of authoritarian rule, in the so-called new nations that achieved independence after World War II or after the cold war, contested narratives of the past are indispensable to any position one takes on the boundaries of sovereignty. This can refer concretely to disputed borders, or more broadly to the question of who belongs to the nation, who does not, and who should rule.

How do nationalists use history "to constitute collective loyalties, legitimize governments, mobilize and inspire people to fight, kill, and die for their country"? Ronald Grigor Suny takes up this question in his wide-ranging essay on the struggles to establish hegemonic narratives of unitary nations in the republics that gained independence after the collapse of the Soviet Union in 1991. Shedding discredited communist ideology and the legacy of Russian and Soviet imperial rule, elites adopted ethnic nationalism as an argument to support their claim to run the state and to define those who might enjoy full membership. To many observers and to the partisans of nationalist movements, this was merely the flourishing of eternal nations liberated from Soviet repression. But as Suny shows, exclusionary narratives of ancient and continuous national identity obscured centuries of experience in the Caucasus, where a shared regional culture and a "polyglot, migrating population" had made the current capitals of Azerbaijan and Georgia into "models of interethnic cohabition." Nationalists who invoked heroic narratives to support their cause did not appreciate the irony of the debt they owed to the USSR, since one unintended consequence of Soviet nationality and modernization policies had been to draw administrative boundaries that artificially hardened fluid ethnic distinctions. In Kazakhstan, historians contributed to the work of forging a new nation by eliminating the multi-ethnic aspects of the Kazakh past and promoting an ethnic Kazakh claim to territorial control. In Armenia, romantic essentialist claims to authenticity and territorial rights reach back much further than the 1915 Turkish genocide that constitutes such an important element of Armenian national consciousness in the diaspora. In the independent republic, Armenian nationalists today vehemently reject proposals—such as one put forth by Suny himself during a visit there—to understand nationality as a combination of historical traditions and subjective will and to emphasize the potentially inclusive character of cultures in the Caucasus.

Subho Basu and Suranjan Das analyze the production of nationalism through narratives of the past on the Indian subcontinent since independence in 1947. They present examples of Hindu and Muslim nationalists wielding partisan histories through the decades before and after independence to seek legitimacy and attract adherents. Basu and Das point out that these historical narratives did not emerge in a vacuum. Rather, they followed a disciplinary tradition established during colonial rule, when the British presented India as "a complex mosaic of static, unchanging, and conflicting well-defined ethnic communities," among which Britain served as a neutral umpire. This colonial project of control survives into the contemporary era of religious nationalist ideology, taking the form of Hindu Rashtra (Hindu state) in India and Islamic fundamentalism in Pakistan, each of them suppressing historical periods of tolerance and syncretism to create an unbroken narrative of purity in states that actually contain great diversity. The extent of constraints on historical debate in Pakistan, the least democratic of the countries discussed in this book, where the questioning of certain founding myths and heroes is banned by law, undergirds our hypothesis that democracies foster more open debate about the past.

The assertion of sovereignty by one group over an area of diverse population on the basis of selective historical claims is also at the heart of the enduring conflict between Israelis and Palestinians, writes Ilan Pappe in his contribution. A popular founding myth of Israel, expressed in slogans such as "making the desert bloom" or "a land without a people for a people without a land," posited the absence or irrelevance of an indigenous population. This provides a good example of what scholars mean when they say that collective memory is a process not only of remembering but also of forgetting. Arabs living in the territory Israelis now claimed were driven away in what Pappe calls the successful ethnic cleansing campaign of 1948. Since then, the facts of 1948 have been sustained in the individual memories of Palestinian refugees holding "property deeds, faded photographs and keys to homes they can no longer return to," and were confirmed by recent scholarship of the so-called revisionist Israeli historians. But an official Israeli historical narrative that avoids the events of 1948 has contributed to the failure of every attempt to negotiate a peace settlement. Reintegrating the neglected history of the ethnic cleansing campaign into the peace process would allow for restitution to Palestinian refugees, in the form of a right of return or financial compensation. In the theory put forward by Howard Zehr, this would enable restorative rather than retributive justice.[11] The partisan histories that substitute for an accurate account that might better enable Israel to come to terms with its past, however, block an understanding of this central Palestinian concern. Pappe joins Edward Said in accompanying this call with the appeal that Palestinians demonstrate their understanding of the significance of memories of the Holocaust to Israeli Jews. Such a twin development in historical empathy may be indispensable to any possible reconciliation.

Nationalism plays a destructive role in most of these accounts, but this need not be a universal law. In Nigeria, a nation fractured by sectional differences, nationalist interpretations of history are part of a project of establishing unity by promoting African or Nigerian traditions, as opposed to colonial or tribal ones, thereby, it is hoped, increasing the viability of the nation-state. As Toyin Falola says, "the colonial library slandered Africa" by presenting Africans as divided and incapable of self-government. Although Europeans created a state called Nigeria, the nation called Nigeria—the "imagined community" that, once established, provides legitimacy to the state and can produce collective allegiance to a central government—must be created through Nigerian narratives of history.[12] In this sense, nationalism can be a positive force. But what should the narrative say? And how does one handle ethnic and religious divisions in developing a national history? There is no consensus about the past in Nigeria, but there are many partisans of rival interpretations. Falola describes the principal categories of historical interpretation that have been used to try to gain broad political support. In the north, political leaders among the Hausa–Fulani appeal for unity based on Islamic tradition. In the south, Yoruba politicians invoke a mythical father figure, Oduduwa, to argue that all Yoruba share a common descent and should belong to the same political party. In the east, the Igbo tend to emphasize another aspect of the past—their widespread suffering during the Biafran War of the late 1960s—to argue that this victimization requires compensation in the form of political power. Falola concludes that in spite of a panoply of attempts by intellectuals and educators to draw on the past for arguments in favor of certain forms of governance, given the challenges of unifying a plural society, discourse on Nigerian national history will continue to be partisan and fragmented.

In exploring the function of partisan histories in political conflicts in the United States, there are many possible subjects one could consider. Campaigns for restitution and redress for past injustices draw the national spotlight from time to time, from the successful efforts of Japanese Americans to gain acknowledgment of their unlawful incarceration during World War II to the campaign for reparations from government and corporations for the descendants of slaves and ongoing struggles by Native Americans to obtain, if not compensation for something that cannot be restored, then various forms of restitution that address acute symbolic grievances (such as the return of human remains held in museums) or provide material relief to make daily life more tolerable. Partisan histories feature in these and in an array of other disputes. Selective symbolic imagery from the Revolution, the civil war, Reconstruction, and World War II plays a role in appeals to party loyalty and clashes over civil liberties, race issues, even tax policy. Legal battles over many subjects are resolved by the courts through a particular interpretation of what the "founding fathers" or "framers of the Constitution" intended.

We have chosen to focus here on the Vietnam War, not because it is more important than all other possible topics, but because it offers a useful example of how disputes over the past can be urgently present in national politics. Rival interpretations of the U.S. intervention in Vietnam 30 years after its conclusion still are very much a part of how Americans make—and justify—decisions on foreign policy.

Patrick Hagopian begins his essay by calling attention to the suppleness of interpretations of the past in providing lessons for the present. Vietnam and Munich regularly arise as historical analogies in contemporary debates over taking military action. The conventional wisdom about the lesson of the 1938 Munich agreement, which granted Adolf Hitler a third of Czechoslovakian territory in exchange for a promise that he would not ask for more, is that negotiating with dictators only emboldens them, and one must intervene early to avoid catastrophic wars. There is no such consensus about the meaning of the Vietnam War. That disastrous episode might teach that there are limits to American power, that one should avoid distant interventions in internal conflicts at the risk of getting caught in a quagmire. To supporters of military intervention, however, the lesson of Vietnam is that one must intervene more forcefully from the start to win a war, not proceed in stages of escalation. Those rival interpretations have been played out in the United States during debates over military intervention abroad from the 1970s into the twenty-first century.

As Hagopian demonstrates, and as we have seen in the other democracies, control over historical interpretation is quite diffuse, and cannot simply be asserted by government. When President Ronald Reagan put forth the "noble cause" interpretation of the Vietnam War in the context of increasing intervention in Central America in the early 1980s, his narrative strategy rallied his conservative supporters but alienated his opponents and much of the public, who feared that if Reagan thought the war in Vietnam had been a good thing, he might try to have another one in Nicaragua or El Salvador. Comparable difficulties faced subsequent administrations with regard to the Gulf War of 1991, intervention in Somalia and former Yugoslavia in the 1990s, and in the war in Iraq in 2003. In each case, competing interpretations of Munich and especially of Vietnam were wielded in public debate and invoked in private deliberations over whether or not to commit U.S. forces.

Hagopian agrees with Ernest May that foreign policy makers often invoke historical analogies simplistically, seldom considering ways in which the comparison might be misleading.[13] In his essay, Hagopian draws on the analysis of Yuen Foong Khong[14] to examine the different ways politicians and their advisers use historical examples: "do they use them *heuristically* as analytical exercises to help them to make decisions; do they use them *didactically* as rhetorical devices to explain their decisions and persuade others; or do they use them *cosmetically* to dignify their decisions after the fact when they write their memoirs, giving their actions a learned appearance by showing how they were informed by historical knowledge?" The same question should be asked about the eight other cases in this book, indeed, about the use of history in political conflicts worldwide, whether these conflicts revolve around national identity and sovereignty, restitution, electoral politics, or foreign policy. The reader may determine that, like the rest of us, policy makers perform all three actions at once: they are greatly influenced in their thinking by narratives about the past that they find compelling; they deploy these narratives to draw support and undercut the opposition; and they turn to interpretations of the past as an unparalleled source of legitimacy on weighty questions.

It may be appropriate to amend Hegel's insight that the only thing we ever learn from history is that we never learn from history. We can be confident that history will continue to be deployed for political purposes, if often with more passion than learning. This collection of essays should encourage a skeptical posture toward political conflicts about the meaning of the past, so that claims of historical justification are not left unexamined.

Notes

1. Peter Novick, *That Noble Dream: The "Objectivity Question" and the American Historical Profession* (New York: Cambridge University Press, 1988).
2. Eric Hobsbawm, *On History* (London: Abacus, 1990), 166.
3. A good introduction to this field can be gained by perusing the journal *History and Memory: Studies in Representations of the Past*, published semiannually since 1989.
4. Maurice Halbwachs, *The Collective Memory* (New York: Harper & Row, 1980, originally published in 1925).
5. Pierre Nora, ed., *Realms of Memory: Rethinking the French Past* (New York: Columbia University Press, 1996–1998, originally published in 1984).
6. Patrick Geary, *Phantoms of Remembrance: Memory and Oblivion at the End of the First Millennium* (Princeton: Princeton University Press, 1994), 12.
7. See A. Dirk Moses, "Coming to Terms with Genocidal Pasts in Comparative Perspective: Germany and Australia," *Aboriginal History*, 25 (2001), 91–115.
8. John Torpey, "Introduction," in Torpey, ed., *Politics and the Past: On Repairing Historical Injustices* (Lanham, MD: Rowman and Littlefield, 2003), 3.
9. Priscilla Hayner, *Unspeakable Truths: Confronting State Terror and Atrocity* (New York: Routledge, 2002), 25.
10. Juan E. Méndez, "Latin American Experiences of Accountability," in Ifi Amadiume and Abdullahi An-Na'im, *The Politics of Memory: Truth, Healing, and Social Justice* (London: Zed Books, 2000), 127.
11. Howard Zehr, *Changing Lenses; A New Focus for Crime and Justice* (Waterloo, Ontario: Herald Press, 1990).
12. Benedict Anderson, *Imagined Communities: Reflections on the Origin and Spread of Nationalism* (New York: Verso, 1983).
13. Ernest R. May, *The "Lessons" of the Past: The Use and Misuse of History in American Foreign Policy* (New York: Oxford University Press, 1973).
14. Yuen Foong Khong, *Analogies at War: Korea, Munich, Dien Bien Phu and the Vietnam Decisions of 1965* (Princeton: Princeton University Press, 1992).

Suggestions for Further Reading

Barkan, Elazar, *The Guilt of Nations: Restitution and Negotiating Historical Injustices* (New York: Norton Press, 2000).

Berg, Manfred and Schäfer, Bernd, eds., *Historical Justice in International Perspective: How Societies are Trying to Right the Wrongs of the Past* (Cambridge, MA: Cambridge University Press, 2005).

Hein, Laura and Selden, Mark, *Censoring History: Citizenship and Memory in Japan, Germany, and the United States* (Armonk, NY: Sharpe, 2000).

Lorey, David E. and Beezley, William H., *Genocide, Collective Violence, and Popular Memory: The Politics of Remembrance in the Twentieth Century* (Wilmington, DE: Scholarly Resources, Inc., 2002).

Minow, Martha, *Between Vengeance and Forgiveness: Facing History after Genocide and Mass Violence* (Boston: Beacon Press, 1998).

Müller, Jan-Werner, ed., *Memory and Power in Post-War Europe: Studies in the Presence of the Past* (Cambridge, MA: Cambridge University Press, 2002).

Paris, Erna, *Long Shadows: Truth, Lies and History* (New York and London: Bloomsbury Publishers, 2000).

PART 1

New Regimes

CHAPTER 1

The Past in the Politics of Divided and Unified Germany

Andrew H. Beattie

The legacies and memories of the past always inform political decisionmaking in the present, but the tumultuous nature of Germany's twentieth-century history has focused considerable attention on the issue of "coming to terms" with the Nazi past (*Vergangenheitsbewältigung*). This is a slippery term that encompasses specific attempts to bring Nazi perpetrators to account and seek justice for their victims, as well as a more general effort to face up to and remember the 12-year period of Nazi rule under Adolf Hitler (1933–1945). The highly emotional and morally charged debate about the adequacy of this process has often hindered, and even substituted for sober analysis of the past's role in Germany's postwar politics.

In Germany since 1945, the issue has been not so much *whether* the past should be remembered, but which aspects of the past gain prominence and become significant politically. Contrary to the common assumption that the Federal Republic of Germany (FRG, or West Germany) was "silent" about Nazism in the 1950s, there was then considerable discussion of certain elements of the Nazi past, particularly the war and its consequences. The emphasis, however, was on Germans' suffering rather than the suffering Germans inflicted on others. Later, the focus shifted more toward German responsibility for Nazi crimes. Yet the Nazi Third Reich was not the only past that played an important role in the politics of the FRG and the German Democratic Republic (GDR, or East Germany). Also at issue in the late 1940s and 1950s were the interwar period of weak democratic rule known as the Weimar Republic (1919–1933), where the causes of the Nazis' rise to power were sought, and the period of postwar occupation (1945–1949), when its consequences were felt most immediately. From the 1960s and especially the late 1970s, the focus, particularly in the FRG, moved away from avoiding a repetition of the collapse of the Weimar Republic and mitigating the war's consequences for the German population, toward developing an appropriate response to the Holocaust in particular. Although never forgotten, addressing German suffering and avoiding Weimar's fate increasingly took second place to memories of the Holocaust.[1]

In addition to being remembered selectively, historical episodes are also interpreted in varying ways, and each variation potentially has political consequences. The "lessons" that German political figures drew from the past were dependent on their interpretations of German history. Generally speaking, the Left understood Nazism as a species of "fascism," an outgrowth of capitalist imperialism. Capitalism therefore had to be replaced with an "anti-fascist" socialism. Conservatives and liberals, by contrast, regarded Nazism as a form of "totalitarianism," of which communism was another variant, and to which the appropriate response was the construction of an "anti-totalitarian" democracy. Such were the competing answers to the Nazi past.

These two mutually exclusive doctrines of antifascism and antitotalitarianism followed very different trajectories in the two Germanys. Antifascism became official doctrine in communist East Germany. In West Germany, the two approaches coexisted, their relative influence rising and falling over time. While antitotalitarianism had the better of antifascism in the 1950s and early 1960s, the late 1960s and 1970s saw a revival of antifascist positions among the West German New Left. By the 1980s, an uneasy stalemate emerged.

Since the unification of the two Germanys in 1990, a combination of continuity and change has characterized the role of the past(s) in German politics. Unification led to a diversification of historical reference points. It elicited, for example, echoes of the previous unification of Germany in 1871. More importantly, with the passing of the postwar era into history in 1990, the number of "German pasts" increased to include those of East and West Germany, and partisan versions of these pasts increasingly informed political decisionmaking and legitimation-strategies. The West German "success story" has motivated and legitimized a remarkable degree of continuity with the old Federal Republic across the caesura of unification, while the East German "failure" has been used to delegitimize alternatives to that western path.

As the Third Reich has receded further in time, its direct political relevance has lessened considerably, yet the major constraints on, as well as justifications for German foreign policies in the 1990s and early 2000s have still concerned the Nazi legacy. The Nazi past and Germans' efforts to face up to it have remained important for political legitimation strategies and continued to dominate German historical memory and identity construction. The fears of many commentators that unification would lead to the weakening of the influence of the Nazi past in the present have thus not been proved correct.[2]

Partisan Histories in Divided Germany

Neither East nor West German politics can be understood outside the context of the cold war and global competition between the United States and the USSR. Both Germanys must be seen as creatures of their occupying powers, as rivals and competitors with each other, and as alternative answers to the Nazi disaster. Attitudes to the past were thus only one of several determinants of German political developments. Yet, even with this caveat, the recent past was of central importance to postwar German politics in numerous ways. Fundamentally, the main goals of German policy—of ending the occupation, preventing or overcoming division, achieving national unity, and regaining full national sovereignty—necessitated "overcoming" the legacy of the recent past.

The two German states' legitimation strategies and their relations with each other were grounded in highly partisan histories. Both sought to draw legitimacy from their disavowal of National Socialism. Consequently, confrontation and competition between them were often couched in terms of alleged dubious continuities with, and successfully learned "lessons" of the past. East and West utilized the doctrines of antifascism and antitotalitarianism respectively to interpret the past, legitimize the present, and delegitimize each other.[3] West German leaders accused the communist GDR of being a totalitarian dictatorship fundamentally similar to the Third Reich, pointing to political persecution, a lack of democracy, and single party rule to support this claim, in contrast to the democratic system established in the West. The GDR leadership, meanwhile, accused the Federal Republic of being a militaristic, revanchist state, whose elites were ridden with former Nazis, indicating lingering and incipient fascism, in contrast to the antifascist order established in the East. The Socialist Unity Party (SED), as the East German Communist Party was called, released "Brown Books" listing West German elites with compromised pasts, and the West responded in kind, pointing to numerous former Nazis in East German institutions. Regardless of the considerable degree of truth, misrepresentation, and hypocrisy on both sides, these were cold war partisan histories of the most basic kind, blatantly instrumentalizing the past to discredit the rival state.

Just as important to the states' claims to legitimacy were their claims to be the true and only proper representative of the German nation and its genuine interests, and their denials of similar status to their rivals. These, too, were based on highly partisan versions of history. After an initial period in which anti-Nazi resisters were regarded in the general population as national traitors, the Federal Republic's posturing as the true representative of the German people was increasingly founded on the legacy of the failed plot by senior military figures to assassinate Hitler on July 20, 1944. This act of bravery, it was argued, rescued the honor of the German people, insulating them against accusations from abroad (often more imagined than real) of collective guilt in Nazi crimes. The resisters' legacy—the good Germany—was realized, of course, in the Federal Republic.[4] In contrast, the FRG ignored communist resistance and portrayed socialism as a Soviet import and the GDR as the artificial product of Soviet occupation. West German authorities persisted into the 1960s in referring to the GDR as the zone of occupation rather than a state. While the degree to which the SED regime rested on Soviet bayonets is open to debate, this western depiction of the GDR ignored a number of facts: that communism and socialism had strong German roots; that, in the immediate aftermath of the war, many Germans and most political parties supported some version of socialism as the only answer to fascism; that communists played productive roles in western state and local governments before being marginalized and persecuted in the developing cold war; and, finally, that the FRG itself was largely the creation of the western occupying powers.[5]

East Germany also claimed to be the proper, and best, representative of the nation, dividing German history into "a progressive line which allegedly culminated in the GDR and a reactionary line which conversely found its pinnacle in the FRG."[6]

It emphasized the strong tradition of German communism, portraying itself as fulfilling the legacy of Marx, Engels, Rosa Luxemburg, and other figures of the German labor movement. In the late 1970s and 1980s, the SED altered its approach, seeking to integrate previously frowned upon historical figures, such as Luther, Bismarck, and Frederick the Great, into its pantheon, leading to a competition with the West over ownership of the inheritance of the national past.[7]

Within the confines of their cold war blocs, the two Germanys' international relations were also closely related to the past, which provided both motivation and justification for political action. Both states' foreign and security policies were driven by a desire to be respected as responsible, peaceful international citizens, and to achieve good neighborly relations with other European states. War should "never again" emanate from German soil. In this vein, West German foreign policy displayed an "ideology of reticence" about asserting German national interests, suggesting not only that policy makers sought to avoid a repetition (or fear among their neighbors of a repetition) of past aggression, but also that foreign policy was a vehicle for making amends for past abuses.[8] Making up for the past was one of the central driving forces behind the successful integration of West Germany into the Atlantic community, achieved by chancellor and leader of the Christian Democratic Union (CDU) Konrad Adenauer (1949–1963), and it lies behind considerable German support for international cooperation, multilateralism, and European integration, even to this day.[9] Importantly, Adenauer also initiated compensation payments to survivors of the Holocaust and sought conciliation with Israel. Like other policies designed to come to terms with the past, this step, which was controversial among the electorate and unpopular even within the government, certainly had an element of self-interest, for reconciliation with Jews and their supporters, it was hoped, would advance the cause of western integration and the ultimate goal of regaining sovereignty.[10] Coming to terms with the past was also among the key motivations of a new West German policy toward Eastern Europe (*Ostpolitik*), initiated by the leader of the Social Democratic Party (SPD) Willy Brandt as foreign minister (1966–1969) and then chancellor (1969–1974), which turned attention to the conduct of the World War II on the eastern front and sought reconciliation with the Eastern European victims of German aggression. This was a profound change from earlier years when the claims of German expellees from former eastern territories had prevented recognition of Eastern European borders.[11] On a more popular level, public opposition to the rearmament of the two Germanys in the 1950s was founded to a considerable extent in a pacifism born of the traumatic experiences of two world wars, a legacy with which the advocates of rearmament and military activity have had to contend ever since.[12]

The international context provides illustrations not just of how the past informed and was instrumentalized in political conflicts, but also of how political developments influenced attitudes toward the past and its place in the present. The cold war served to marginalize self-critical memories within Germany about German crimes, and delay or inhibit the pursuit of justice for those crimes. The superpowers needed the support of their German client states and were complicit in allowing a high degree of continuity among compromised technical, scientific, and military elites, as attentions shifted from the past Nazi danger to the present cold war threat.[13]

Partisan Histories in East Germany

The "founding fathers" of both states had been politically active in the Weimar Republic and remained profoundly influenced by their experiences, although their differing interpretations led to divergent responses. Communists stressed the significance of the early Weimar period and the failures and "betrayal" of the "incomplete" German revolution of 1918–1919, which supposedly enabled fascism's rise. The lessons drawn from this included a determination not to repeat past errors by leaving the reactionary elites—who had helped the Nazis to power—intact, or by allowing the working-class parties to remain divided and thus ineffectual against the forces of reaction.[14] A widespread, if arbitrary denazification process therefore took place during the "anti-fascist democratic upheaval" (1945–1949) in the Soviet Zone of Occupation (which became the GDR in 1949). Large landholders and factory owners were expropriated, and the SPD was pushed into a merger with the Communist Party, creating the SED.

With the foundation of the GDR, the necessary steps to overcome the past had apparently been taken: Nazis were removed from power, the remnants of feudalism were crushed, political power for the party of the working class was secured, and friendly relations were established with the Soviet Union, whose role in defeating the fascist armies was stressed. Moreover, the abolition of capitalism was well under way, and thus the social and economic basis for a potential fascist revival was destroyed. Although there was ongoing need for vigilance against reactionary elements and the supposed neofascist threat from the West, SED leaders argued that priority henceforth could be given to the construction of socialism, rather than continuing efforts to come to terms with the past. For the SED, history's role in informing specific political decisions had ceased.

Antifascism, however, continued to be utilized to embarrass the West and legitimize SED rule generally. That many of the GDR's leading figures had been persecuted by the Nazis and survived in exile, like SED first secretary Walter Ulbricht (1949–1971), or in prisons and concentration camps, like SED general secretary Erich Honecker (1971–1989), gave the regime's antifascist posturing a degree of plausibility. It also successfully limited the development of political opposition even into the 1980s, as many intellectuals, like writer Christa Wolf, felt that they could not oppose those who had fought the Nazis.[15] But the designation of the Berlin Wall—constructed in 1961 to prevent ever more East Germans from abandoning socialism for the capitalist West—as the "anti-fascist protective wall" signaled that official GDR antifascism was degenerating into a hollow ideology for maintaining power. East Berlin consistently regarded the Holocaust as peripheral to fascism (which was regarded as a class rather than a racial phenomenon), and treated its Jewish victims with disdain relative to the much-lauded communist "anti-fascist resistance fighters." It was left to concerned individual citizens to investigate the Holocaust, which they did increasingly in the 1980s.[16] By 1988–1989, opposition groups, comprising younger people free of the inhibitions felt by Wolf's generation, sought to reclaim German historical figures like Rosa Luxemburg from the clutches of official propaganda, turning the regime's antifascist, democratic, and socialist tenets against itself.[17] The strictures of official antifascism were broken, and soon after losing control over the past, the regime also lost political power.

Partisan Histories in West Germany

Interpretations of history were more diverse in the democratic West than the dictatorial East. Various approaches to the past and visions for the future had to compete with one another in a relatively free public sphere. Under the influence of the cold war, an initial antifascist consensus that had existed in the immediate aftermath of the war—encompassing communists, Social Democrats, liberals and conservatives—gave way to a conservative "anti-totalitarian consensus" that was effectively reduced to anticommunism, and was also directed against Social Democrats.[18] Confronting the past for its own sake, or that of justice, took second place to unspecific exhortations to stand up to the totalitarian threat this time around.

The totalitarian reading of the recent past had profound effects on West German politics. In contrast to East German communists' emphasis on the revolution of 1918–1919, in West Germany the focus was more on avoiding a recurrence of the collapse of democracy at Weimar's end. According to the totalitarian reading, Weimar's fragile parliamentary democracy had succumbed to attack from communists as well as Nazis, and new institutions had to be strengthened relative to those of Weimar, in order to defend themselves better against all forms of extremism and totalitarianism in the present. The liberal and conservative architects of the Basic Law (*Grundgesetz*), as the West German constitution was called, therefore developed a "corrective" to the Weimar constitution.[19] It guaranteed human and civil rights, enhanced the power of the chancellor, weakened the powers of the president, introduced constructive votes of no-confidence in the chancellor, anchored the role of political parties in the con- stitution, introduced a five percent hurdle for gaining entry to the Federal Parliament (*Bundestag*) to exclude extremist parties and avoid atomization, and established a strong constitutional court and a federal, decentralized system of representative, rather than participatory government. These were all responses to Weimar's failure.

As well as seeking to prevent another collapse of parliamentary democracy, the *Grundgesetz* had redemptive qualities and constituted a "counter-constitution in oppo- sition to the Third Reich," declaring, for example, human life to be sacred.[20] (This step appears very progressive, but it hindered campaigns for the legalization of euthanasia and abortion in West Germany and continues to complicate German responses to issues such as genetic engineering.)[21] In acknowledgment of those who had fled persecution in Nazi Germany and those who had granted them protection, the *Grundgesetz* also granted asylum to anyone suffering political persecution. (In a move widely seen as indicating the diminishing significance of the Nazi past in post-unification Germany, this right to asylum was heavily circumscribed by a change to the *Grundgesetz* in 1993.)[22]

Perceptions of a need to avoid a repetition of Weimar's demise—the "Weimar syndrome"—continued to influence West German political and intellectual life even after the proclamation of the *Grundgesetz*.[23] A concept of "militant democracy" (*wehrhafte Demokratie*) was developed as an attempt to learn from the earlier repub- lic's failure to defend the democratic center against attacks from the political extremes. The state actively fought political radicalism, banning the neo-Nazi Socialist Reich Party and the West German Communist Party in 1951 and 1956 respectively. Even in the 1970s, the shadow of Weimar's collapse informed the state's tough stance against the left-wing terrorism of the Red Army Faction, while the screening of applicants to the public service (including teachers) to keep out left-wing radicals was justified with reference to the undemocratic mindset of Weimar bureaucrats.[24] With the passing years, however, the question of whether the FRG would avoid Weimar's fate could be answered with increasing assurance in the negative.[25] By the 1980s, the "Weimar syndrome" was over, and Weimar was no longer a compelling source of partisan histories.

West German antitotalitarianism emphasized present dangers to democracy rather than the undemocratic past. Indeed, the very *absence* of self-critical discussion of the past played a significant political role. In 1949, many Germans believed the Allies had already dealt with the Nazi past, and Adenauer pursued a set of policies aptly characterized as a combination of "amnesty, integration and distancing": while the Nazi regime was officially condemned, the vast majority of those who had served it were treated with remarkable leniency and integrated into the new state and its booming economy. Not talking too explicitly about German crimes was crucial to a politics of integration that blurred distinctions between perpetrators, bystanders, and victims.[26] Yet the 1950s were not a period of silence or repression of the past, but one of greater concentration on German suffering (whether from Allied bombing, expulsion from Eastern Europe, rape by Soviet soldiers, or as POWs) than on German crimes.[27] Such selective memory produced a range of legislative measures seeking to address German suffering, like the Equalization of Burdens Law that compensated Germans for property lost during the war. These helped to integrate potentially disgruntled sections of the population, albeit at the expense of justice and an explicit confrontation with German responsibility for Nazi crimes.[28]

From the end of the 1950s, a series of events, including incidents of anti-Semitic graffiti in late 1959, the capture and trial of Nazi war criminal Adolf Eichmann in Jerusalem in 1961, and a number of trials in German courts, provoked fresh awareness of German crimes and seemed to suggest that previous efforts to deal with the Nazi past had been insufficient. These events coincided with processes of generational change, as the generation of the "founding fathers" gave way to two younger generations. The first, the "forty-fivers," were born in the 1920s and had few memories of the Weimar period but were deeply influenced by their youthful experiences in the Third Reich and its defeat and collapse. This generation, which includes many of most influential political and intellectual figures in the FRG, pursued a "project of consolidation and reform" of the republic.[29]

Following closely on their heels was a still younger and more radical generation, which constituted the German chapter of the global protest phenomenon of 1968. Born in the 1940s, its members had few if any personal memories of Nazism and were socialized in the early postwar decades. These "sixty-eighters" rejected their parents' claims of innocence and ignorance of Nazi crimes and were highly critical of their obsession with German suffering. They felt deep skepticism toward the postwar order that had been established with so many continuities with the past and believed that fascism lurked in the 1960s present.[30] Despite their youth, the "sixty-eighters" partook in the "Weimar syndrome," seeing in proposed "Emergency Laws," which would give the state sweeping powers in the case of an external or internal threat, a repetition of the self-abnegation of Weimar democracy. They condemned Western anticommunism and sympathized with East German antifascist attacks on the FRG, even if they did not see their dream of socialist revolution realized across the Berlin Wall. For this New Left, coming to terms with the past also meant opposing Israeli Zionism and U.S. imperialism.[31]

A left-wing minority of "forty-fivers," like the philosopher Jürgen Habermas, agreed with many of the criticisms of the "sixty-eighters," even if they did not agree with their unconventional and often violent methods, especially as elements of the New Left descended into terrorism. What they shared with more moderate "sixty-eighters" was a deep skepticism about the German nation as a pre-political community of fate. Whereas in the early postwar years the rise of the Nazis was often depicted as a natural catastrophe or a "spanner in the works" of German history, left-liberal historians of these two postwar generations insisted that Nazism was the result of longer currents, even pathologies, in German history, which was regarded as constituting a special path (Sonderweg) of negative historical development. Indeed, for much of the population, and the Left in particular, awareness of Germans' historical guilt complicated, hindered, if not morally proscribed positive national identification, as exemplified by writer Günter Grass's argument that Auschwitz rendered a unified German nation a moral impossibility, or writer Walter Jens's description of unification as a "dream void of history."[32] The only answer to the Sonderweg seemed to be the championing of universal western values and "post-nationalism," meaning a collective identity based on loyalty to abstract constitutional principles, rather than an emotional attachment to the nation, which had after all produced the gas chambers. The twin projects of post-nationalism and Holocaust memory found their natural political home in the New Left offshoot, "The Greens," but their intellectual and political significance exceeded that party's relatively minor electoral standing, as "sixty-eighters" assumed key roles as teachers, journalists, and other professionals.

Their negative reading of German history did not go unchallenged. After being out of government since 1969, the CDU gained power under Chancellor Helmut Kohl in 1982. Kohl declared a "spiritual-moral turn" that sought to wind back many of the social and cultural transformations that had taken place in the 1960s and 1970s, resurrect 1950s antitotalitarianism, and refocus attention on German suffering. He and his conservative intellectual allies sought the establishment of a (West) German national pride based on postwar economic and political achievements (as did the SED in the GDR in relation to its economic, social, and sporting achievements). Necessarily this project was closely connected to a relativizing approach to the Holocaust, which blurred distinctions between victims and perpetrators and between Germans and their former enemies, and equated Nazi with communist crimes. A counterreaction from liberals and the Left produced the "Historians' Debate" (Historikerstreit) of 1986–1987, which amounted to a contest over the singularity of the Holocaust and its significance for contemporary German politics. Conservatives advocating a positive national identity and playing down the significance and uniqueness of the Holocaust were unable to overcome the resistance of liberals and left-wingers, who insisted on the singularity of German crimes and the consequent impossibility of an uncomplicated, positive national identity.[33] While neither antifascism nor antitotalitarianism could claim victory, a consensus emerged in the political middle ground recognizing the genocide of Europe's Jews as a crime of unprecedented proportions with profound consequences for German identity and politics.

Partisan Histories in Germany Since Unification

The collapse of the SED regime in late 1989 and the unification of East and West Germany in 1990 prompted many questions about the new Germany's future path and the role of the past in German politics. The most common term for the accession of the newly reformed East German states to the Federal Republic, "reunification" (*Wiedervereinigung*), itself raised numerous questions: what exactly was being "re-"unified, to what prior condition was Germany now returning, which former incarnation of German statehood, if any, could serve as a model for the country, and, conversely, which period or entity was now being laid to rest?

At the most obvious level, unification ended the cold war division of the nation, yet the term *Wiedervereinigung* ("wieder" meaning "again" as well as "re-") also recalled the first unification of Germany under Prussian hegemony in 1871 to form the Second Reich. Would the reunified nation amount to a "Fourth Reich," which would (once again) take a nationalist turn and upset the European order, as some, including British prime minister Margaret Thatcher, feared? The immediate post-unification period did see a partial revision of previously overwhelmingly negative attitudes toward the Second Empire (although this tendency dated back to the German–German competition in the 1980s over ownership of the national past), but not even conservative politicians regarded the empire as worthy of uncritical assessment or indeed emulation, and its significance for post-unification politics was minor.[34]

That this past was discussed at all indicates a diversification of historical reference points subsequent to unification. More recently than this discussion in the early 1990s, an apparent crisis in confidence in the German polity, as well as deepening problems facing the welfare state, have led to numerous references to the Weimar past. The consensus, however, is that Germany today is so far removed from Weimar, both temporally and qualitatively, that any parallels drawn between the two are superficial, and those who draw them are generally met with derision.[35]

For German politics, the most significant "new" pasts are those of East and West Germany, which both became history with the passing of the postwar era in 1990, albeit to starkly differing degrees. The significance of these pasts in political decision-making was evident in the debate about whether Bonn or Berlin should be the seat of the government and parliament of the new Germany. While Berlin was the official capital, a decision had to be made about leaving these institutions in Bonn, or moving them to Berlin. A number of present-centered considerations were involved, including the parochial interests of Bonn residents and the costs of moving, but symbolic questions arising from Germany's multiple pasts and their legacies in the present were paramount. Provincial Bonn had come to symbolize the West German "success story," in particular the stability and federal nature of its political system and its successful western integration. Berlin, meanwhile, had been the capital of authoritarian militarist Prussia, the doomed Weimar Republic, as well as the totalitarian Third Reich, and it was from Berlin that Germany had launched two world wars. Many feared that moving back to the metropolis would lead to a more centralist mode of politics and a strident foreign policy oriented more toward Central and Eastern Europe. Yet Berlin also evoked memories of western resistance to communist tyranny (e.g., in the Berlin Blockade of 1948–1949) and provided the ultimate symbol of cold war division and its end: the Berlin Wall. In the present day, Berlin was a microcosm of reunified Germany, facing daily the challenges of uniting the western and eastern halves of the city.

Ultimately, 338 members of the *Bundestag* voted for Berlin, narrowly defeating the 320 who opted for Bonn. The major parties, the Christian Democrats and Social Democrats, were split, and it was the votes of the SED's successor party, the Party of Democratic Socialism (PDS), for Berlin that decided the day. The decision indicated that the burden of Berlin's pre-1945 pasts would not hinder it pursuing a political future, while the present-centered commitment to the "growing together" of the nation motivated many parliamentarians to vote for Berlin.[36] Nevertheless, the sizeable vote for Bonn demonstrated the strong desire among many West Germans not to broach any changes subsequent to unification: the new Germany should fundamentally be an enlarged continuation of the old Federal Republic.[37] The related and ongoing discussion about the designation of unified Germany as the "Berlin Republic," distinct from the Weimar and Bonn republics, demonstrates continuing western reluctance to accept any substantive discontinuity with West Germany.[38] Indeed, the entire, extraordinarily asymmetric process of German unification—amounting to the wholesale transfer of western political, social, and economic institutions and values to the eastern states—implicitly delegitimized the East German past, where everything was held to be bad, while raising the West German past to the status of an almost unquestioned norm, where everything was held to be good.[39] Cold war fronts continued even after the end of the cold war.

The East German and the Nazi Pasts

Following unification, the East German past was frequently referred to as the "second German past," and there was much talk of Germany's "double past." The first term suggested that, hitherto, there had only been one past: the Nazi past. It also provided further indication of the degree to which the old Federal Republic was regarded as continuing, largely unaltered, for no such designation was applied to West German history. More importantly, the "double past" lumped the GDR together with the Third Reich, indicating that the GDR constituted Germany's "second dictatorship," which was deemed to require a second or "double *Vergangenheitsbewältigung*."[40]

The major political role of the GDR past in the post-unification period was its utilization by political parties to discredit their rivals. In March 1992, the *Bundestag* established a Commission of Inquiry for "Working Through the History and Consequences of the SED-Dictatorship in Germany." It denounced the GDR as a totalitarian dictatorship and laid responsibility firmly at the feet of its ruling party, the SED, whose successor, the PDS, was by implication to be denied political respectability and legitimacy, even a political future. According to former Eastern dissident and Commission chair Rainer Eppelmann (CDU): "Through the precise analysis of the totalitarian structures of rule of the SED dictatorship, the Commission of Inquiry was to help in ensuring that those forces that fundamentally organised the oppression of people in the GDR never get a political chance in unified German again." In response, the PDS postured as the defender of the East German past against Western attack, objecting to the instrumentalization of history for political purposes and the demonization of the GDR, pointing to the GDR's "achievements" (its antifascism and social policies), and alluding to "problems" in West German history (such as the persecution of communists and continuities with the Third Reich).[41] The antitotalitarian versus antifascist dichotomy was thus maintained.

Every party except the PDS supported the inquiry's condemnation of GDR communism. Conservative Christian Democrat members of the commission majority, however, indicted not just Soviet-style communism as practiced in the GDR, but any variant of socialist thought and practice, hoping to demonstrate "why the ideology of socialism *had* to lead to a system that was so contemptuous of people."[42] Even notions of a third way between capitalism and socialism were thus to be smeared with the totalitarian brush, which was directed as much against Social Democrats and Greens from the old West German Left, for being too soft on communism, and those East German dissidents who still harbored reform socialist ideas, as against the PDS. In response, while western Social Democrats objected to the reduction of Marxism to its instrumentalization and perversion by the SED, eastern dissidents played down their desire for a reformed GDR.[43] Advocates of socialist alternatives to the Federal Republic were thus confined to the post-communist fringe. There were also ongoing disputes between the SPD and CDU about the extent to which Christian or Social Democrats had resisted or assisted the communists in the establishment and maintenance of the dictatorship.[44] Accusing the other side of co-responsibility for the GDR, while emphasizing one's own resistance and persecution, was good politics.

Anyone who failed adequately to condemn communism and abhor its legacy was regarded as having left the field of democratic (and therefore admissible) political and historical positions. As long as it refused to subscribe to the (otherwise) unanimous reading of the postwar past, the PDS was politically beyond the pale. Yet what would happen if it abandoned its apologias for communist crimes and accepted a more self-critical stance? The 2001 elections to the Berlin House of Assembly demonstrated how partisan histories have been used not just to discredit rivals, but also to legitimize coalition formation. Following the collapse of the SPD–CDU government and failed negotiations over forming a coalition with minor parties, the SPD was left with little option but to seek a coalition with the PDS. To counter claims from outside and within its own party that it was being soft on communism, the SPD leadership pressured the PDS into distancing itself from certain aspects of the past. In a preamble to the coalition agreement, the PDS condemned injustices in the GDR and acknowledged the SED's "ongoing guilt for the persecution of Social Democrats" in particular, adopting a range of positions and language that it had heretofore rejected.[45]

As well as providing ammunition for party-political competition and serving to narrow the field of acceptable political positions and historical interpretations, the process of coming to terms with the East German past was also intended to bring East and West closer together. The notion that dealing with the past should play a central role in promoting "inner unity"—subsequent to the formal political and constitutional unity achieved in 1990—can be seen as recognition of the fundamental importance to the two German states during the period of division of their opposing partisan histories. It also indicated the decisive role attributed to historical consciousness in the creation of national identity. In contrast to a democratic preference for valuing intellectual pluralism, politicians from all parties suggested that only when the divergent views of history were abandoned in favor of a single interpretation of the past would the nation truly be united. The *Bundestag* Commission, therefore, sought the construction of a new, joint understanding of German postwar history that would help overcome lingering mental barriers between East and West. It was common for that history to be cast in terms of a "history of division," which at least in principle should include Germans from both sides of the border (as opposed to an exclusive focus on the history of the GDR, which might not concern westerners).[46]

One would expect, then, the commission to address postwar German history *in toto*, but the bulk of West German history was off limits, and PDS calls for an investigation of both sides of the divided nation were rejected for fear of suggesting that the FRG and the GDR were equally legitimate and their histories equally problematic.[47] Instead of examining East and West German history as two sides of a common past, the inquiry's focus was overwhelmingly directed toward the negative aspects of the GDR, although it also sought to celebrate the legacy of the East German revolution that overthrew the communist regime in 1989–1990.[48] As well as examining its political system and repressive apparatus, a key component of the commission's indictment consisted in the exposure and condemnation of East German antifascism as (and its reduction to) a state-legitimizing doctrine ordered from above. Moreover, through its use of the totalitarian paradigm, the commission implicitly and explicitly compared the GDR with Nazi Germany.[49] It is not without justification that some perceive the return and triumph of the 1950s' cold war totalitarianism.[50]

The extent of the victory of the antitotalitarian Right can be seen in the commission's proclamation of an "anti-totalitarian consensus," that, based on the condemnation of both "totalitarian" regimes, was to be (re-)established as the only legitimate political position available in the new Federal Republic. This amounted to a significant concession from the West German Left in particular, as indicated by the fact that its philosophical leader, Jürgen Habermas, self-critically spoke of the need for an "anti-totalitarian consensus that deserves this name because it is not selective."[51] Yet the equationist Right did not have everything its own way. While accepting the comparison of the GDR and Nazi dictatorships, the Left argued that they could not be equated, let alone played off against one another, and neither could they assume equal status in contemporary Germany. Many moderate conservatives now also regarded the singularity of the Holocaust (both as a historical event and as a factor in the present) as paramount. The commission acknowledged fundamental differences between the Third Reich and the GDR, recognizing that "that which makes the Nazi dictatorship an ineluctable inherited burden for generations to come did not exist in the SED dictatorship."[52]

The decision, taken under the CDU government of Chancellor Kohl, to build a central "Memorial to the Murdered Jews of Europe" near the Brandenburg Gate in the heart of Berlin's new government precinct indicated the extent to which Holocaust memory had broad support across the German political class.[53] The lack of such a national memorial dedicated to either the victims of communism or the German victims of World War II and its aftermath, including the expellees from former eastern territories, demonstrates that Holocaust memory holds a predominant position in contemporary Germany. Yet precisely that predominance continues to be challenged by some on the Right, among them a number of former East German dissidents. Despite the supposed consensus reached in the *Bundestag* Commission, conservative members objected to the continuing dominance of the Holocaust for German memory, identity, and politics, and continued to equate the two forms of "totalitarianism." In a second inquiry, they sought to put the memory of the victims of communism and Nazism on the same footing, particularly at the sites of former concentrations camps that had been used as Soviet internment camps after 1945 and memorials to antifascist resistance in the GDR (and therefore had a double or even triple past). In the end, a compromise formula was adopted, leaving sufficient room for interpretation to please both sides: Nazi crimes were not to be relativized by the debate on communist crimes, while communist crimes were not to be trivialized through reference to Nazi crimes.[54] This position stood in opposition to the Right's equationist tendency to lump Nazi crimes and victims together with communist crimes and victims. Yet it also implied a rejection of the Left's privileging of the former at the expense of the latter, and the commission explicitly opposed any hierarchies among victim groups.[55] Opposition to the predominant position of Jewish and other victims of the Nazis remains, and those who advocate "more memory" of communism and German suffering have considerable political support.[56] The debate continues.

The West German and the Nazi Pasts

West German history provided normative guidance for the unification process generally and was largely spared the critical examination the GDR past received in the 1990s. It was not completely exempted, however. Unification brought about a curious reversal of positions in relation to the old Federal Republic. In the 1980s, conservative views of the FRG were overwhelmingly positive, except that leftist influence was decried, while many liberals and the Left were critical of its lingering social conservatism and the continuing reluctance of the Right to accept the singularity of the Nazi past. Faced with the return of the nation-state in 1990, however, the West German Left discovered just how much it had grown to appreciate (and influence) West German liberal democracy and feared that unification would bring renewed nationalism and a host of other ills.[57]

The conservative reading of the FRG remained largely positive, influencing the overall unification process and providing considerable support for Bonn in the capital debate. However, the "New Right" and many conservatives, including Chancellor Kohl and the parliamentary leader of the CDU, Wolfgang Schäuble, believed the time had come to make some significant changes in particular areas, including public memory, national identity, and foreign policy. Once again, this meant undoing "1968." The ideological disputes of the 1970s and 1980s, which, it was felt, the Left had won, had to be re-contested.[58] In the Commission of Inquiry, conservatives used the history of the FRG to extend praise or condemnation, respectively, for tough or soft stances on communism and national division. The West German Left (especially the New Left) was accused of accepting, or at least failing to condemn national division and socialist dictatorship, thus forsaking the national brethren in the East in favor of post-national, antifascist, and socialist sympathies. "What discouragement," said Schäuble, "it must have meant back then for the people in the GDR . . . that GDR socialism found so much sympathetic understanding in some political and in many so-called intellectual circles" in the West.[59] Those who had come to accept division as permanent and abandoned the goal of unification were severely reprimanded, while those who had clung to the dim hope of an end to division (like Kohl, the "Chancellor of Unity") were praised as genuine patriots. Thus, in contradistinction to the post-national tendencies of the Left, the formerly divided and now reunited nation was affirmed and normalized, while political parties fought over whose policy had facilitated eventual unification.[60] Beyond the commission, concerted campaigns have sought to embarrass former "sixty-eighters" such as Joschka Fischer (Greens) over their earlier opposition to West German democracy or their connections with left-wing terrorists.[61] Such attacks, though, have had little electoral impact. With the victory of the SPD and Green coalition in federal elections in 1998 and 2002, the "sixty-eighters" achieved political power at the highest level.

Foreign policy debates were also linked to re-readings of the German pasts. The left-liberal understanding of Germany's history between 1871 and 1945 as a disastrous *Sonderweg* underwent considerable revision after unification, leading to a more positive approach to the "Second Reich."[62] Taking this transformation of the *Sonderweg* notion a step further, a number of conservatives and members of the New Right argued that the Federal Republic itself should now be seen as the deviation from the norm, rather than the normative answer to previous derailment. Its apologetic bearing and multilateralist, restrained foreign policy could now be replaced by a more assertive, "self-confident nation."[63] Chancellor Kohl's hesitance in acknowledging Germany's border with Poland in 1990 and Germany's recognition of the breakaway Croatian Republic in 1991, contrary to EU policy, seemed almost to confirm the fears among many of Germany's neighbors that unified Germany would pursue its national interests in a more unilateral fashion and reassert its historical interests in Eastern and Southeastern Europe.[64]

However, in part to assuage these concerns, the Kohl government intensified its commitment to furthering the project of European integration, and the reluctance of both the German government and the wider population to consider military participation in international conflicts, such as the Gulf War of 1991, suggested that the "bad" old Germany was not in fact returning, and that the "constraining power of collective memory" persisted.[65] Nevertheless, that power was diminished by the fact that both sides of the debate over military participation instrumentalized the past, drawing "lessons" that seemed increasingly arbitrary. In the debate over German participation in the Kosovo crisis at the end of the 1990s, opponents of German participation argued that the lesson of the past was "never again war!" while those who favored joining the NATO action to stop ethnic cleansing by Serbia, including Foreign Minister Joschka Fischer, argued that the appropriate lesson was "never again Auschwitz!" The latter response suggested that Germany, if anything, had a particular obligation to prevent genocide and humanitarian disaster and was justified in using military intervention to this end.[66]

German foreign policy decisionmaking today clearly is based on assessments of national interests, electoral possibilities, and normative concerns in relation to particular issues. The German government's opposition in 2002–2003 to the impending war against Iraq had as much to do with concerns about the war's potential consequences for the Middle East and an objection to the U.S. government's new doctrine of unilateral, preemptive strikes, as it did with any specific legacies of the German past. The popularity of the position, which helped return the SPD–Green government to power in the elections held in 2002, was not only due in part to the "culture of antimilitarism" that developed in Germany after its bitter experiences of warfare in the first half of the twentieth century, but it was also due to an assessment of the case for war on its merits, popular anti-Americanism, and the appeal of national decisiveness and rejection of subservience to the United States.[67] Today, "learning the lessons" of the past is thus but one of a number of determinants and aims of German foreign policy, alongside the pursuit of national interests. A preference for nonmilitary, multilateral solutions is clearly a legacy of the past, but these central tenets of Germany's postwar policy are increasingly disconnected from specific attitudes to the past and are becoming mere vehicles for the expression of national interests.[68]

Conclusion

Post-unification Germany has witnessed a diversification of historical reference points, as previous incarnations of German nation- and statehood have appeared politically relevant at particular moments. While the pasts of the Second Reich and the Weimar Republic have re-asserted themselves only fleetingly, the twin postwar pasts have played crucial political roles. At the same time, as the Nazi past has receded further into history, the "lessons" it implied in postwar Germany have become increasingly irrelevant. Partisan histories of Nazism were always utilized for present-day purposes, but in the post-Wall era such instrumentalization appears increasingly blatant and unhelpful, its implications rather arbitrary.

Yet Nazism still provides a fertile source for legitimation strategies. Unification did not bring the marginalization of the Nazi past, as many feared. On the contrary, the Berlin Holocaust memorial and a series of contemporary debates—whether on the wartime crimes of the German army, the bombing of German cities, or the commemoration of May 8, 1945 as a day of defeat or liberation—testify to the salience of the Nazi past. In part, this is due to the development of an international Holocaust memory culture. In part it is due directly to the strength of the "Holocaust identity" that developed in West Germany: the Holocaust's perceived singularity and its ongoing significance are recognized by a broad spectrum from the Left to the moderate Right, so that more right-wing efforts to displace the memory of Nazi crimes from the center of German politics have been defeated.[69] In large part, however, it is because the "new" German pasts tended to concern the Nazi legacy, albeit at one remove. As we have seen, efforts to settle accounts with communism actually served to revive debate about East and West German attempts to "come to terms" with the Nazi burden, as conservatives sought to demolish East German antifascism as a mere myth. The main thrust of the delegitimation of the SED regime, meanwhile, consisted in comparing and equating it with its Nazi predecessor. Condemning communism by association with Nazism only strengthened the centrality (and negative normativeness) of the latter; thus the very efforts to shift Nazism from its central position have only served to cement it there.

The implications for German identity are complex. For all that divided them, the interpretations of the immediate past underlying the two doctrines of antitotalitarianism and antifascism, in whose name the postwar states were founded, were overwhelmingly negative. The recent past provided little practical or intellectual guidance, beyond a catalog of what had to be avoided; learning the "lessons" and avoiding the "mistakes" of the past were the appropriate responses.[70] The immediate need for such guidance faded as the postwar regimes became firmly established. Subsequent generations, however, far from exhibiting less critical views, developed even more negative approaches to the national past, seeing pre-1933 periods, indeed the entire course of German history, as thoroughly compromised by the Holocaust. For many on the Left in particular, a positive national identity was impossible. Many on the Right, in turn, objected to the centrality of the Holocaust in German political culture, precisely because it hindered a more positive relationship to the nation and its history.[71]

Since unification, there have been signs of change. For the Center-Right, positive identification with the nation no longer seems to be incompatible with *Vergangenheitsbewältigung*, while for the Center-Left the Holocaust no longer proscribes national identification completely. Indeed, many evince a certain pride based precisely on the strenuous and ongoing German efforts to deal with the "double past," and there are other positive legacies and precedents on which to draw. According to the likes of both former Chancellor Kohl and his successor Gerhard Schröder (SPD), Germans can take pride in the achievements of economic recovery and parliamentary democracy in West Germany, and the regaining of national unity by peaceful means. Closely related to the question of pride is the question of German "normalcy": can Germany, finally, regard itself, and behave as a "normal" nation (whatever that may be)? The discussion of foreign policy above suggests that in pursuing its national interests, German policy makers are certainly heading in that direction.

Yet a compelling vision of national identity or a consensus about national "normalcy" is still to emerge. Even the minimal "anti-totalitarian consensus" of *opposition* to totalitarianism and extremism is beset with problems of application that make questionable whether it can form the basis of a stable democratic collective identity. The highly partisan and inconsistent approaches to the West and East German pasts, moreover, have alienated East and West Germans as much as they have integrated them.[72] The occasional call to celebrate the legacy of the East German revolution of 1989 has not been heard amongst the general condemnation of the GDR, and for many Germans unification provides little to celebrate among massive unemployment, depopulation, and mounting reconstruction costs in the East. Political divisions also remain. The renewed "anti-totalitarian consensus" was founded against the post–Communist Left, while Social Democratic support for such a consensus sits oddly with continuing skepticism about the totalitarian paradigm and the latter's application to the GDR. Meanwhile, conservative dissatisfaction with the hierarchical nature of the consensus—with Nazism and the Holocaust still ascendant—has not disappeared.[73] Similarly, it is unlikely that those who demand that greater attention be paid to German suffering will be satisfied by well-meaning suggestions that it, too, be allocated a subordinate position in a hierarchy with Jewish victims and German perpetrators at the top, for it is precisely that hierarchy to which they object.[74] The diversification of German pasts has loosened the strictures of cold war interpretation, but for each past there are multiple partisan histories that continue to divide rather than unite.

Notes

1. Robert G. Moeller, "Sinking Ships, the Lost *Heimat* and Broken Taboos: Günter Grass and the Politics of Memory in Contemporary Germany," *Contemporary European History*, 12: 2 (2003), 147–181.

2. See Ian Kershaw, *Germany's Present, Germany's Past* (London: Institute of Germanic Studies, 1992).

3. Antonia Grunenberg, "Antitotalitarianism Versus Antifascism—Two Legacies of the Past in Germany," *German Politics and Society*, 15: 2 (1997), 76–90, here 77.

4. Jürgen Danyel, ed., *Die geteilte Vergangenheit: Zum Umgang mit Nationalsozialismus und Widerstand in beiden deutschen Staaten* (Berlin: Akademie Verlag, 1995).

5. Patrick Major, *The Death of the KPD: Communism and Anti-Communism in West Germany, 1945–1956* (Oxford: Clarendon Press, 1997), 257–293; Eric D. Weitz, "The Ever-Present Other: Communism in the Making of West Germany," in Hannah Schissler, ed., *The Miracle Years: A Cultural History of West Germany, 1949–1968* (Princeton: Princeton University Press, 2001), 219–232, here 228.

6. Stefan Berger, *The Search for Normality: National Identity and Historical Consciousness in Germany since 1800* (Providence: Berghahn, 1997), 70.

7. Mary Fulbrook, *German National Identity after the Holocaust* (Oxford: Polity, 1999), 86, 91–92, 130–139.

8. Andrei S. Markovits and Simon Reich, *The German Predicament: Memory and Power in the New Europe* (Ithaca: Cornell University Press, 1997), xii.

9. Jeffrey Herf, *Divided Memory: The Nazi Past in the Two Germanys* (Cambridge, MA: Harvard University Press, 1997), 220.

10. Herf, *Divided Memory*, 286–287; Fulbrook, *German National Identity*, 65–66.

11. Robert G. Moeller, "What Has 'Coming to Terms with the Past' Meant in Post–World War II Germany?" *Central European History*, 35: 2 (2002), 223–256, here 241–242; Herf, *Divided Memory*, 344.

12. Corey Ross, " 'Protecting the accomplishments of socialism'? The (re)militarisation of life in the German Democratic Republic," in Patrick Major and Jonathan Osmond, eds., *The Workers' and Peasants' State: Communism and Society in East Germany under Ulbricht 1945–1971* (Manchester: Manchester University Press, 2002), 78–93, here 81–82; Michael Geyer, "Cold War Angst: The Case of West German Opposition to Rearmament and Nuclear Weapons," in Schissler, *The Miracle Years*, 376–408.

13. Herf, *Divided Memory*, 261–262.

14. Martin Sabrow, "Kampfplatz Weimar: DDR-Geschichtsschreibung im Konflikt von Erfahrung, Politik und Wissenschaft," in Heinrich August Winkler, ed., *Weimar im Widerstreit: Deutungen der ersten deutschen Republik im geteilten Deutschland* (Munich: R. Oldenbourg, 2002), 163–184.

15. Mark R. Thompson, "Reluctant Revolutionaries: Anti-Fascism and the East German Opposition," *German Politics*, 8: 1 (1999), 40–65.

16. Angelika Timm, *Jewish Claims Against East Germany: Moral Obligations and Pragmatic Policy* (Budapest: Central European University Press, 1997), 50; Joachim Käppner, *Erstarrte Geschichte: Faschismus und Holocaust im Spiegel der Geschichtswissenschaft und Geschichtspropaganda der DDR* (Hamburg: Ergebnisse Verlag, 1999).

17. Laurence H. McFalls, *Communism's Collapse, Democracy's Demise? The Cultural Context and Consequences of the East German Revolution* (New York: New York University Press, 1995), 96.

18. Sigrid Meuschel, "Legitimationsstrategien in der DDR und in der Bundesrepublik," in Christoph Kleßmann, Hans Misselwitz, and Günter Wichert, eds., *Deutsche*

Vergangenheiten—eine gemeinsame Herausforderung: Der schwierige Umgang mit der doppelten Nachkriegsgeschichte (Berlin: Christoph Links, 1999), 115–127, here 117.

19. Meuschel, "Legitimationsstrategien," 121.
20. Bernd Faulenbach, "Historical Foundations of the Berlin Republic," in Dieter Dettke, ed., *The Spirit of the Berlin Republic* (New York: Berghahn, 2003), 9–23, here 12.
21. Eric Langenbacher, "Changing Memory Regimes in Contemporary Germany?" *German Politics and Society*, 21: 2 (2003), 46–68, here 47; Manfred D. Laubichler, "Frankenstein in the Land of *Dichter and Denker*," *Science*, 286 (December 3, 1999), 1859–1860.
22. See Andreas Huyssen, "Nation, Race, and Imigration: German Identities After Unification," in *Twighlight Memories: Marking Time in a Culture of Amnesia* (New York: Routledge, 1995), 67–84; Jürgen Habermas, "The Asylum Debate (Paris Lecture, January 14, 1993)," in *The Past as Future* (Lincoln: University of Nebraska Press, 1994), 121–141.
23. A. Dirk Moses, "The 'Weimar Syndrome' in the Federal Republic of Germany," in Stephan Loos and Holger Zaborowski, eds., *Leben, Tod und Entscheidung: Studien zur Geistesgeschichte der Weimarer Republik* (Berlin: Duncker & Humblot, 2003), 187–207, here 191.
24. Moeller, "What has 'Coming to Terms with the Past' Meant," 229; A. Dirk Moses, "The State and the Student Movement in West Germany, 1967–1977," in Gerard J. de Groot, ed., *Student Protest: The Sixties and After* (London: Longman, 1998), 139–149.
25. William E. Paterson, "From the Bonn to the Berlin Republic," *German Politics*, 9: 1 (2000), 23–40, here 25.
26. Norbert Frei, *Adenauer's Germany and the Nazi Past: the Politics of Amnesty and Integration* (New York: Columbia University Press, 2002).
27. Robert G. Moeller, *War Stories: The Search for a Usable Past in the Federal Republic of Germany* (Berkeley: University of California Press, 2001).
28. Michael L. Hughes, *Shouldering the Burdens of Defeat: West Germany and the Reconstruction of Social Justice* (Chapel Hill: University of North Carolina Press, 1999).
29. A. Dirk Moses, "The Forty-Fivers: A Generation Between Fascism and Democracy," *German Politics and Society*, 17: 1 (1999), 94–126, here 117.
30. Moeller, "Sinking Ships," 158, 161–162; Moses, "The Forty-Fivers," 119.
31. Christhard Hoffmann, "The Dilemmas of Commemoration," *German Politics and Society*, 17: 3 (1999), 1–8; Herf, *Divided Memory*, 349.
32. On Grass, Jan-Werner Müller, *Another Country: German Intellectuals, Unification and National Identity* (New Haven: Yale University Press, 2000), 64–89; Jens cited in Lothar Probst, "Deutsche Vergangenheiten—Deutschlands Zukunft: Eine Diagnose intellektueller Kontroversen nach der Wiedervereinigung," *Deutschland Archiv*, 27: 2 (1994), 173–180, here 173.
33. Charles S. Maier, *The Unmasterable Past: History, Holocaust and German National Identity* (Cambridge, MA: Harvard University Press, 1988).
34. Bernd Weisbrod, "German Unification and the National Paradigm," *German History*, 14: 2 (1996), 193–203; Lutz Niethammer, "Geht der deutsche Sonderweg weiter?," in Antonia Grunenberg, ed., *Welche Geschichte wählen wir?* (Hamburg: Junius, 1992), 23–54; Dettke, *The Spirit of the Berlin Republic*, 3.
35. Paterson, "From the Bonn to the Berlin Republic," 25; Volker Wörl, "Ein abwegiger Vergleich: Oskar Lafontaines Parallel mit der Politik des Reichskanzlers geht an den Realitäten vorbei," *Süddeutsche Zeitung*, November 22, 2002; Michael Stürmer, ". . . aber nicht Weimar: Der Vergleich der Berliner und der Weimarer Republik ist so plakativ wie falsch," *Die Welt*, November 21, 2002.
36. Kurt Sontheimer, "Berlin schafft keine neue Republik—und sie bewegt sich doch," *Aus Politik und Zeitgeschichte*, 51: 1–2 (2001), 3–5; Stuart Parkes, *Understanding Contemporary Germany* (London: Routledge, 1997), 54.

37. Heinrich August Winkler, "Rebuilding of a Nation: The Germans Before and After Unification," *Daedulus*, 123: 1 (1994), 107–127, here 117.
38. Paterson, "From the Bonn to the Berlin Republic," 23.
39. Helga A. Welsh, "The Divided Past and the Difficulties of German Unification," *German Politics and Society*, 30 (1993), 75–86, here 78–81.
40. Christoph Kleßmann, "Das Problem der doppelten 'Vergangenheitsbewältigung'," *Die Neue Gesellschaft*, 38: 12 (1991), 1099–1105; Konrad H. Jarausch, "A Double Burden: The Politics of the Past and German Identity," in Jörn Leonhard and Lothar Funk, eds., *Ten Years of German Unification: Transfer, Transformation, Incorporation?* (Birmingham: Birmingham University Press, 2002), 98–114.
41. Deutscher Bundestag, ed., *Materialien der Enquete-Kommission "Aufarbeitung von Geschichte und Folgen der SED-Diktatur in Deutschland" (12. Wahlperiode des Deutschen Bundestages)* I (Frankfurt: Suhrkamp, 1995), 182, 11–17, 87, 89, 680–692.
42. Emphasis added. Alfred J. Gertler, "Gemeinsame Herausforderung: Bundestag will Enquete-Kommission über das DDR-Unrecht einsetzen," *Flensburger Tageblatt*, February 13, 1992; Roswitha Wisniewski, "Das Bildungssystem der DDR und sein Vermächtnis," in Peter Barker, ed., *The GDR and Its History: Rückblick und Revision* (Amsterdam: Rodopi, 2000), 129–143, here 131–132.
43. Deutscher Bundestag, *Materialien der Enquete-Kommission*, 268–272, 585–605; Eckhard Jesse, "Oppositionelle Bestrebungen in der DDR der achtziger Jahre: Dominanz des Dritten Weges?," in Karl Eckart, Jens Hacker, and Siegfried Mampel, eds., *Wiedervereinigung Deutschlands: Festschrift zum 20jährigen Bestehen der Gesellschaft für Deutschlandforschung* (Berlin: Duncker & Humblot, 1998), 89–101.
44. Deutscher Bundestag, *Materialien der Enquete-Kommission*, 35, 65, 90–91, 110, 127, 233–247.
45. "Koalitionsvereinbarung zwischen SPD und PDS für die Legislaturperiode 2001–2006," January 8, 2002, http://www.pds-berlin.de/politik/dok/koalentw_p.html
46. Deutscher Bundestag, *Materialien der Enquete-Kommission*, 154, 182, 184, 186, 187, 192.
47. As Rainer Eppelmann stressed in an interview with the author, Berlin, August 27, 2001.
48. Deutscher Bundestag, *Materialien der Enquete-Kommission*, 745.
49. Ibid., 190, 192, 215, 278–281.
50. Bill Niven, *Facing the Nazi Past: United Germany and the Legacy of the Third Reich* (London, Routledge, 2002), 61.
51. Deutscher Bundestag, *Materialien der Enquete-Kommission*, IX, 690.
52. Ibid., I, 744–745.
53. Langenbacher, "Changing Memory Regimes," 58; Niven, *Facing the Nazi Past*, 194–232.
54. Deutscher Bundestag, ed., *Materialien der Enquete-Kommission "Überwindung der Folgen der SED-Diktatur im Prozeß der deutschen Einheit,"* I (Frankfurt: Suhrkamp, 1999), 614.
55. Ibid., I, 617.
56. Deutscher Bundestag, "Antrag der Abgeordneten Günter Nooke et al., 'Förderung von Gedenkstätten zur Diktaturgeschichte in Deutschland—Gesamtkonzept für ein würdiges Gedenken aller Opfer der beiden deutschen Diktaturen'," *Drucksache*, 15/1874 (Berlin: H. Heenemann, 2004), 6.
57. Daniel Levy, "The Future of the Past: Historiographical Disputes and Competing Memories in Germany and Israel," *History and Theory*, 38: 1 (1999), 51–66, here 66.
58. Jacob Heilbrunn, "Germany's New Right," *Foreign Affairs*, 75: 6 (1996), 80–98.
59. Deutscher Bundestag, *Materialien der Enquete-Kommission "Aufarbeitung,"* I, 65.
60. Ibid., I, 36, 41.
61. M. Anne Sa'adah, " 'Ein Staatsman mit Geschichte': Joschka Fischer's German Past," *German Politics and Society*, 19: 3 (2001), 56–79.

62. Konrad H. Jarausch, "Normalisierung oder Re-Nationalisierung? Zur Umdeutung der deutschen Vergangenheit," *Geschichte und Gesellschaft*, 21 (1995), 571–584, here 577; Berger, *The Search for Normality*, 201, 176.

63. Heimo Schilk and Ulrich Schacht, eds., *Die selbstbewußte Nation: "Anschwellender Bocksgesang" und weitere Beiträge zu einer deutschen Debatte* (Berlin: Ullstein, 1994).

64. Beverly Crawford, "Explaining Defection from International Cooperation: Germany's Unilateral Recognition of Croatia," *World Politics*, 48: 4 (1996), 482–521; Ann L. Phillips, "The Politics of Reconciliation Revisited: Germany and East-Central Europe," *World Affairs*, 163: 4 (2001), 171–191.

65. Markovits and Reich, *The German Predicament*, 6.

66. Jeffrey K. Olick and Daniel Levy, "Collective Memory and Cultural Constraint: Holocaust Myth and Rationality in German Politics," *American Sociological Review*, 62: 6 (1997), 921–936, here 933; Hans Christoph Buch, "Falsche Lehren aus Europas Historie," *Die Welt*, April 9, 2003.

67. Thomas U. Berger, *Cultures of Antimilitarism: National Security in Germany and Japan* (Baltimore: Johns Hopkins University Press, 1998); William M. Chandler, "Foreign and European Policy Issues in the 2002 Bundestag Elections," *German Politics and Society*, 21: 1 (2003), 161–176.

68. Tuomas Forsberg, "The Debate over Germany's Normality: a Normal German Debate?" in Howard Williams, Colin Wight, and Norbert Kapferer, eds., *Political Thought and German Reunification: The New German Ideology?* (Basingstoke: Macmillan, 2000), 139–155, here 152; Rainer Baumann, "The Transformation of German Multilateralism: Changes in the Foreign Policy Discourse since Unification," *German Politics and Society*, 20: 4 (2002), 1–26.

69. Jeffrey C. Alexander, "On the Social Construction of Moral Universals: The 'Holocaust' from War Crime to Trauma Drama," *European Journal of Social Theory*, 5: 1 (2002), 5–85; Bernhard Giesen, *Intellectuals and the Nation: Collective Identity in a German Axial Age* (Cambridge: Cambridge University Press, 1998), 152.

70. Grunenberg, *Welche Geschichte wählen wir?*, 17.

71. Müller, *Another Country*, 45–63.

72. Laurence McFalls, "Political Culture and Political Change in Eastern Germany," *German Politics and Society*, 20: 2 (2002), 75–92, here 90.

73. Bernd Faulenbach, "Die Arbeit der Enquete-Kommissionen und die Geschichtsdebatte in Deutschland seit 1989," in Barker, *The GDR and Its History*, 21–33, here 31; Jörg Lau, "Fatales Abwägen: Das Holocaust-Gedenken gerät in Konkurrenz zur Aufarbeitung des Stalinismus," *Die Zeit*, online 7 (2004).

74. Aleida Assmann, "Funke einer gesamtgesellschaftlichen Erregung: Eine Frage der Hierarchie: Leid und Schuld sind in der deutschen Erinnerung keineswegs so unvereinbar wie es scheint," *Frankfurter Rundschau*, online, February 3, 2004.

Suggestions for Further Reading

Frei, Norbert, *Adenauer's Germany and the Nazi Past: the Politics of Amnesty and Integration* (New York: Columbia University Press, 2002).

Fulbrook, Mary, *German National Identity after the Holocaust* (Oxford: Polity, 1999).

Garton Ash, Timothy, *In Europe's Name: Germany and the Divided Continent* (London: Vintage, 1994).

Herf, Jeffrey, *Divided Memory: The Nazi Past in the Two Germanys* (Cambridge, MA.: Harvard University Press, 1997).

Marcuse, Harold, *Legacies of Dachau: The Uses and Abuses of a Concentration Camp, 1933–2001* (Cambridge: Cambridge University Press, 2001).

Maier, Charles S., *The Unmasterable Past: History, Holocaust and German National Identity* (Cambridge, MA: Harvard University Press, 1988).

Markovits, Andrei S. and Reich, Simon, *The German Predicament: Memory and Power in the New Europe* (Ithaca: Cornell University Press, 1997).

McAdams, A. James, *Judging the Past in Unified Germany* (Cambridge: Cambridge University Press, 2001).

Niven, Bill, *Facing the Nazi Past: United Germany and the Legacy of the Third Reich* (London: Routledge, 2002).

Roberts, Geoffrey K., *Party Politics in the New Germany* (London: Pinter, 1997).

CHAPTER 2

Apologizing for the Past between Japan and Korea

Alexis Dudden

A transnational movement of national apology has become a key feature of world politics since the late twentieth century. Leaders of states and other groups seeking formal recognition around the globe now routinely remember past atrocities in public in order to inscribe their legitimacy anew.[1] As they apologize for wrongs that the international community perceives as abnormal to its collective sense of self, they become articulators of new official histories. Such apology then defines participating nations as "normal states" and "members in good standing" of the global community.

The extent to which government bodies—including the Vatican with its observer status at the United Nations—engage in apologetic politics depends in part on their protagonists' aspirations to power in the international system. Within the same week in July 2002, for example, even two *non*members of the international community of states—the so-called rogue-state North Korea and the Irish Republican Army—made international apologies, in an effort on their part to gain some sort of legitimacy in the world community.[2]

Usually, civic groups acting with and on behalf of victims of state-sponsored atrocities pit themselves against their own or foreign governments to bring about state apologies in the first place. Victims, activists, and their sympathizers, therefore, regard official apology as a worthy goal; when a government apologizes in some form, many count it as an achievement. Such apologies fuel demands for more examination of past wrongdoings. In recent years, however, government leaders have slowly co-opted the substance of apology to their own ends, using the past in the present to benefit the state's national interests. While many continue to think of apology purely in terms of a victory for victims, states have managed to find value in apology, too.

In 1985, German president Richard von Weizsäcker escalated the stakes of international apology for the past with his highly publicized statements concerning

Germany's former Nazi regime. As a result of Weizsäcker's actions and words, as well as the German government's numerous official acts of restitution and commemoration over half a century, many have come to view Germany as the "apologetic model." In contrast, many regard Japan as the world's unapologetic renegade, with most discussions of Japan and apology focusing on Japan's inability to apologize, or the inadequacy of Japan's apologies compared with Germany's behavior.[3] This essay will take a different approach in order to bring into relief the historical depth of Japan's engagement in apology politics, arguing that Japan *is* an apologetic nation according to international standards, yet those standards fail to achieve victims' aims.

Although discussions of how Japan should come to terms with its wartime behavior in Asia began almost as soon as Emperor Hirohito surrendered in August 1945, Japan's 1982 whitewashing of the term "aggression" into "advancement" in school textbooks inaugurated the postwar era's most intense debates over how Japan would account for the nation's attempt to conquer Asia during the first half of the century. The era of high economic growth in Japan increased the available public conceptual space for national self-reflection, a trend that gained more momentum with Hirohito's death in 1989, spurring on even broader participation in the debate. Now entering its third decade, Japanese apologetics is a discipline in its own right. Bookstores have devoted sections to "war responsibility," journals specialize in the issue, and the controversy has generated a discourse so familiar that catch phrases stand in place of entire discussions.[4] Even the most cautious political observers define Japan's "textbook problem"—meaning the issue of what events from the war are included in school books in Japan's government-approved education system—as one of the most volatile issues in regional diplomacy, and apology for the past remains the crucial topic in intra-regional affairs.[5]

Emperor Hirohito's death coincided, moreover, with widespread democratization trends throughout the Asian-Pacific nations of Japan's former empire and encouraged thousands of long-ignored and denied voices to come forward to tell their histories. Japan's victims and their supporters around the world and in Japan took advantage of this freer atmosphere to challenge nationally narrated "truths" of Japan's twentieth century.[6] Well-known examples include the former sex slaves of the Japanese military, forcibly conscripted soldiers, and enslaved laborers. Each of these confronted Japan's popular modern myths concerning its so-called war of "liberation" in Asia in the 1930s and 1940s. During the 1990s, sometimes purposefully, and sometimes unintentionally, these people engaged their claims for redress from Japan with a larger, global trend of apology and reconciliation causes, giving rise to what even the movement's most ardent detractors acknowledge as a decade of "apologetic diplomacy."[7]

Meanwhile, the 1990s saw an end to Japan's era of so-called miracle economics. Thus, leaders now anxiously debate Japan's "national interests" in light of their aspirations to have Japan considered a "normal state" in the international system. Many in Japan define "normal" as a nation with a permanent UN Security Council seat and an active military; both of these are costly to maintain.[8] Though they are often ignored, for the country's leaders, Japan's official statements of apology have consequently become deeply intertwined with national interests.[9] Japan's conservative ruling party, in particular, has veered toward what many refer to as the pragmatic approach, which argues that it is in Japan's national interests to apologize for "the past" because doing so affirms Japan's ties with its Asian neighbors.[10] Put differently, doing so also fosters unencumbered economic exchange in the region.

The spate of apologies in the 1990s revealed a hierarchy of terms in the discourse that surrounds reparations and compensation monies: "sorrow" but not "apology"; "regret" but not "condolences." From a legal standpoint, using or not using the term "apology" makes states more or less vulnerable to claims for compensation.[11] Nevertheless, we can learn more about how apologetic politics falls short of satisfying victims' demands by considering all these various terms together, as elements of what I call the discourse of sorrow and regret. States participating in the international system—particularly democratic ones—now commonly rely on this shared vocabulary to incorporate victims' histories into new, "future-facing" versions of themselves. The recent history and structure of official apology demand attention, therefore, because state leaders now manipulate this discourse at the expense of the dignity that victims seek for themselves.

"Future-Facing" Rhetoric on "The Unhappy Past"

Japan's "future-facing" apologetic stance in the Asian-Pacific region began with the normalization and trade agreements that the Japanese government signed after 1945 with the nation's former colonies. It intensified dramatically in the 1990s, as state leaders entrenched this global medium of exchange among "normal states" as an increasingly common, domestic and diplomatic element of governance. They did so in statements that more openly mentioned Japan's colonization of various parts of Asia during the first half of the twentieth century, reversing long-standing practices of purposeful forgetting. As Norma Field has eloquently observed, "What had been official blasphemy seemed to become, overnight, commonsense."[12]

With several notable exceptions, Japanese politicians of all parties now typically speak in the "future-facing" rhetoric of "sorrow and regret" for "the past" that defines the national interests approach to apology.[13] It is common to dismiss this discourse out of hand as lacking in real meaning. This opinion is not misplaced, but it traps those who are sympathetic to it. For many victims, no apology will ever suffice. Yet, many victims and their supporters have collectively determined that the "future-facing" terms of "sorrow and regret" are fake. Thus is created the problem of needing someone to decide when an official apology is "real." Moreover, the "real" versus "fake" standoff has caused many scholars and observers to believe that Japan is the world's unapologetic outcast without fully considering the international historical milieu that concerns Japanese officials.

Japan's problems with apology are not unique to its relations with Korea, and it would be misplaced to argue that one Asian-Pacific country's movement for redress from Japan is more important than another's. Japan's postcolonial relations with Korea afford, however, an unusually rich history for demonstrating the state's determination not to let the problems of the past undermine the present. In no other aspect of Japan's foreign relations have the political techniques of apology been so well developed.

After 13 years of protracted negotiations, in June 1965 Japanese and South Korean diplomats normalized relations. This treaty was Japan's first state-level, post–Peace

Treaty settlement in Asia to address Japan's colonial era. On the eve of normalization, Japanese officials made statements that shed light on Japan's practice of apologetic politics.[14]

At present, the expression "unhappy past" has frozen into an oblique signifier for describing Japan's colonial period in Korea (1905–1945). When South Korean President Chun Doo Hwan visited Japan in 1984, Emperor Hirohito for the first time publicly stated his personal "regret" over the "unhappy past" the countries shared.[15] Labels such as these for the colonial era have a special ability not to name anything, which rightly angers those seeking recognition of specific histories. Such labels can be traced back to the negotiations leading to the 1965 treaty: it was in fact U.S. ambassador to Japan Edwin O. Reischauer who encouraged the insertion of this apologetic formula into Japanese–Korean negotiations.

Reischauer strongly felt that the Japanese government should address Korean resentment over colonization; recently declassified Johnson administration papers reveal his personal involvement in Japan's statements.[16] At heart, Reischauer espoused the then-prevailing pragmatic view of the U.S. administration that normalization was urgently needed. Robert Komer, of the National Security Council staff, succinctly explained why in a memo to National Security Advisor McGeorge Bundy:

> Top priority in NE Asia today is *ROK/Jap settlement*. This could mean so much more in the way of long-term US dollar savings than a troop cut that there's no comparison. We're still spending over *$300 million* a year on 20 million ROKs, with really no end in sight. So we've got to find someone to share the long-term burden, and it's logically the Japs. Settlement would pump $.6 to 1 billion of public and private funds into ROK, with more later.[17]

Despite Tokyo's reluctance, Reischauer held firm to his position that Japan should make an official statement about the past. In September 1964, he wrote a memo to Secretary of State Dean Rusk acknowledging that "clear Japanese apology for their colonial oppression of Korea in past" was difficult because "Japanese officials and public simply do not feel they owe any apology to Koreans."[18] In November, he sent a telegram to Rusk to report on his private breakfast meeting with Foreign Minister Shiina Etsusaburo during which he urged Japan to make "some sort of apology to Koreans for colonial past." When Shiina's secretary suggested that the foreign minister's upcoming visit to Korea would come "as close to expression of apology as was feasible," Reischauer responded favorably, yet added that "some sort of forward-looking statement about turning backs on past unhappy history . . . might assuage Koreans' feelings without irritating Japanese public."[19]

Early the following winter, in February 1965, Shiina visited Seoul for several days. Socialists in Japan protested the Liberal Democrats' (LDP) decision to normalize relations with South Korea alone by organizing a no-confidence vote in the Diet, and students in both Tokyo and Seoul held mass demonstrations protesting the move to exclude North Korea from the settlement. Japanese prime minister Sato Eisaku and South Korean president Park Chung Hee, however, were determined to make Shiina's visit a success at all costs. When Japan's foreign minister arrived at Seoul's Kimpo airport, he immediately declared Japan's "regret" for "the unhappy period" the countries shared.[20] Shiina told waiting reporters, "I believe that both countries should reflect deeply on the truly regrettable circumstances of the unhappy period in the midst of our nations' long history . . . It is in the hopes of both countries that we establish future-facing permanent and friendly relations on which we can build a new respectful and prosperous history." The countries had not yet established diplomatic relations, and this moment marked the first Japanese official public statement in Korea—South or North—about the colonial era. *Asahi* newspaper special correspondent Imazu Hiroshi noted Shiina's statement as highly significant, and he conveyed his own hopes for improvement in his article quoting Shiina.[21] At the same time, James C. Thomson, Jr., of Lyndon B. Johnson's National Security Council Staff, condescendingly remarked that "Shiina came as close as a Japanese can to apologizing for Japan's sins, and everyone—including State—is thoroughly pleased."[22]

In pragmatic terms, Reischauer's apologetic notion of "unhappy history" neatly froze the past into an indeterminate time period for which no one was to blame—the perfect solution for a national interests approach to an uncomfortable past. And yet, the callousness of such a formulation is clear: democratic states that rely on such "future-facing" rhetoric humiliate survivors of atrocity—whether their own nationals or foreigners—by saying, in effect, "That was then, this is now, you don't matter to our future, and therefore your past must be swallowed for the benefit of our present." The state does not necessarily deny the past, but victims are left with no option for protest; they run along a Möbius strip of the state's creation in courts that ultimately remain indifferent to their claims.[23]

Before considering the trajectory of Shiina's airport declaration, it is worth noting that Japan's records of its policymaking process at this time are unusually difficult to obtain. When the U.S. State Department's Office of the Historian recently published its own records of U.S. Foreign Relations during Lyndon B. Johnson's presidency, it acknowledged an uncommon editorial decision in the preface to its planned-for two-volume Korea/Japan series:

> The most substantive declassification problem arose not among documents selected for publication on Korea, but for those selected for Japan. It was (our) unanimous view . . . that given the number and significance of documents selected for publication in the Japan compilation that must still remain classified, the Japan part of the volume did not constitute a "thorough and accurate, and reliable documentary record of major United States foreign policy decisions." . . . Part 2 of this volume on Japan will not be printed until it meets these standards.[24]

The documents "must still remain classified" because the Japanese foreign ministry will not agree to their public release. The U.S. State Department's Office of the Historian is appealing the Japanese government's decision, yet acknowledges that, "We win some of those appeals and some we lose . . . It can take many years to settle."[25]

Japan's refusal to participate in the timely declassification of documents is not the only occasion when the international community has tolerated the perspective of Japanese officials and their apologists toward their nation's former colonies. Ironically, however, in 1965 the Japanese government actually wanted publicity for its negotiations with Korea and encouraged the major Japanese newspapers to report on the foreign minister's trip to Seoul as well as the talks later that year in Tokyo. As former *Asahi* correspondent Imazu Hiroshi made clear to me, journalists at the time had access to Japanese officials in ways that young reporters only dream about today.[26] Imazu's full-page article "Four Days in Seoul" stands as a rare piece of reportage that blends private and public insights into a vivid and informed history of the event.[27] Imazu believed that both governments negotiated in good faith, yet he also paid particular attention to student protests as well as to police efforts to prevent protestors' access to the diplomats. Imazu underscored that the South Korean people were by no means in unanimous accord with their government's decision to resume relations with Japan.

Imazu drew attention to the problem that lay ahead for Koreans who sought state-level redress from Japan. Noting that his Korean counterpart at the *Donga Ilbo* emphasized a student banner that read "Stop New Colonialism!!", Imazu observed, "The reality of colonial rule in the past is enough by itself to make [Koreans'] unease quickly turn to fear, which, in turn, becomes conviction. If we [Japanese] are really in earnest about bringing our two countries closer together, then we have to look beyond money and goods and think about 'problems of the heart.' This above all was what I learned during my brief stay in Seoul."[28] Imazu's commentary foreshadowed how the official formula of "regret" for "unhappy history" might allow Japanese to think that the problem was solved. While some Koreans—businessmen and some government officials—would profit directly from normalized relations, others would later discover the gap between formula and meaning to be unbearably great.

How, then, did Reischauer and Shiina's 1965 terms shape Japan's politics of apology? Both as a diplomat and more famously as a historian, Edwin O. Reischauer's belief in the unbounded benefits of modernity led him to encourage a future-oriented approach to the past. The discourse of "sorrow and regret" emerged as the most powerful apologetic technique in play. Japanese leaders finally solidified their "future-facing" stance in these terms during the 1990s, when similar political apologies flourished worldwide. Repeatedly during the intervening decades, Japanese and South Korean diplomats and politicians made statements that "regretfully" dissolved "the past" into indeterminate declarations concerning the nations' future together. Surely even the most basic psychoanalysis would describe this as denial.

Reischauer's involvement demonstrated U.S. pressure regarding this particular Japanese policy, yet both Japan and South Korea have subsequently chosen to maneuver within the boundaries of this formula. As a result, Japan's official apologetic techniques have, until now, helped define South Korea as the internationally recognized government on the Korean peninsula.[29] This circumstance, of course, did not exist during the era of "unhappy history" when there were no separate countries, only "Korea." The 1965 Normalization Treaty between Japan and South Korea identified the latter as the "lawful Government" on the Korean peninsula. Notably, until the historic head-of-state meeting in September 2002 between Tokyo and Pyongyang, Japan's official proclamations of "regret" for the "unhappy past" referred solely to South and not North Korea, thus quietly hardening South Korea's privileged position into stone. The international community further exacerbated this selectivity, referring to "Japanese–Korean relations," on the one hand, and "Japanese–North Korean" ones on the other. In short, only the South has existed in Japan's apologetic imagination for the past several decades. Given the importance of that relationship, the substance of any future normalization between Japan and North Korea will likely maintain the national interests involved in that hierarchy, benefiting Tokyo first (as did normalization with the South in 1965), then Seoul, and then, finally, Pyongyang.

Simply put, the "future-facing" nature of Japanese–South Korean relations has polarized rather than expanded perspectives of the past. The intensity of claims against Japan in the 1990s only further entrenched this polarization, as the officials involved worked within the parameters of the existing discourse. More than 30 years after Shiina's statements at Kimpo airport, in October 1998, Japanese Prime Minister Obuchi Keizo issued Japan's first written declaration of "regret" for "the past" to South Korean president Kim Dae-jung.[30] There were no active verbs in the groundbreaking joint declaration that heralded a "new partnership" between the countries. When national interests are at stake, state leaders sidestep grammatical constructions that might require someone to take responsibility for history. The heads of state subsequently celebrated the six-month anniversary of their declaration in March 1999 with an unusual live television broadcast. Because no journalist raised the dreaded issue of the past, neither did Obuchi or Kim. Instead, both men spoke of the dawn of a "new history," prodding business and military leaders to forge ahead under the rubric of "cultural sharing," an expression that eerily evoked the policies of the 1920s.[31]

Since 1965, the South Korean government's acquiescence to Japan's apologetic practices has weakened its own citizens' demands that Japan recognize their histories. As was evident in his "new partnership" declaration (1998), even South Korea's most internationally recognized advocate of free expression, former president Kim Dae-jung, relied on the frozen forms of international apology. When South Korea's current and likewise activist president Roh Moo-hyun made his inaugural trip to Tokyo in June 2003, it seemed only a matter of course that he, too, agreed to "face the future" with both the prime minister and emperor of Japan.[32]

Kim's participation in the politics of apology with Japan reveals the grip such practices hold on international relations, and suggests that a critical rethinking of these forms is necessary. In a 1998 interview, Kim told Okamoto Atsushi, editor of the Japanese opinion journal *Sekai*, that although the Japanese government was responsible for compensating survivors of Japan's sexual slavery, he would refrain from lodging a protest at the United Nations on the surviving women's behalf. "Although Japan's 'comfort women' policy was indeed horrendous," Kim remarked, "what Germany did to Jews was even worse. The German people, however, have recognized this fact and educate their children accordingly."[33]

Kim pointed to the critical need for a long-term commitment in the form of social education. In doing so, he followed the contours of the widely supported "Japan should be more like Germany" approach. Yet, by choosing the safe course of invoking Germany as the model apologizer, Kim eschewed the opportunity that only a head of state attains in apology politics. He was the only one who could meaningfully challenge the official terms of "the past." Ultimately, Kim's explanation devalued the horror that the Korean women endured. He respected the status quo in agreeing to Japan's "regret for the past," and suggested that Japan's state-sponsored atrocities were not as bad as Germany's. Kim stayed within the parameters of official apologetic technique as he described the suffering of Korean victims as less than that of the victims of a more familiar state-sponsored atrocity—the German death camps. In this way, Kim weakened his own citizens' claims to regain their dignity in public before they die.

Sporting Apologies

In June 2002, the jointly hosted Korea–Japan World Cup soccer tournament finally placed in full public view the power and tenacity of Japan's apologetic practice. Notwithstanding the daily rhetoric of how the World Cup was improving Japanese–Korean relations (of course, Japanese–South Korean relations), several events surrounding the spectacle demonstrated that the World Cup was ultimately a very pricey means of reinforcing the status quo.[34]

A year before the games began, relations in the region were at a postwar low; skeptics wondered if the joint venture would happen at all. In August 2001, after months of angry protests from Korean and other Asian leaders over the Japanese government's latest approval of whitewashed textbooks, Prime Minister Koizumi Junichiro visited the notorious Yasukuni Shrine to Japanese war dead.[35] His visit was the first prime ministerial tribute in five years, and provoked official condemnation from South Korean president Kim Dae-jung, huge protests in Seoul, ambassadorial recalls, and demands from around the world for Japan to come to terms with and apologize for its past.[36] Consequently, at the beginning of a long-planned year of joint "cultural exchange" in tandem with the World Cup (the Japan and Korea Foundations respectively spent millions on arts performances and visiting scholar/student programs), the South Korean government banned whole categories of Japanese cultural imports.

Then, all of a sudden, Korean protests against Japan were put aside. The 9/11 attacks on the World Trade Center and the Pentagon in the United States overshadowed this regional battle, and Japan and South Korea redoubled their efforts to "face the future" together. Arguably, the new U.S.-led war on terrorism has secured the Japanese government in its strongest position on apology and South Korea since 1965. Put bluntly, the United States has mobilized its most important ally—Japan—into high gear as the distant military base that the United States has unofficially required of Japan throughout the post-1945 era. In exchange, the United States overlooks Japan's official recalcitrance with regard to "the past" in Asia, and it does so with even greater myopic aplomb than usual.[37]

Nowhere, however, was Japan's U.S.-sanctioned disdain for Asian victims' claims more apparent than in Prime Minister Koizumi's apology tours to China and South Korea in October 2001.[38] Japan's conservative prime minister hopped off to both countries for a day each where he uttered the three long-desired words of reconciliation in short order ("apology," "condolences," and "regret"), and returned immediately to Tokyo without spending the night.[39] Koizumi practiced apology without jet or history lag. He thus reduced the hotly contested issue of Japan's official acknowledgment of responsibility for World War II to its most instrumental level, turning his back on "unhappy history" and placing Japan squarely in a "future-facing" position for a new war.

President George W. Bush had already declared North Korea a member of his infamous "axis of evil" when Koizumi visited Seoul for a second time in March 2002.

Japanese and Korean leaders had scheduled this moment to announce progress on the business world's eagerly anticipated Free Trade Agreement. Although neither leader had reacted favorably to Bush's expression, the U.S.-led war abroad in Afghanistan had become such a powerful feature of daily life in the two countries—both countries' militaries were already involved abroad—that the two leaders could not do enough to reiterate their intention to "face the future" together. Kim told reporters, "This meeting will find its place in history as a building block to the future," and Koizumi echoed, "We will endeavor to make this year pave the way for a bright and forward-facing future."[40] The two leaders, furthermore, made clear to North Korea their resolve to stand squarely with the United States, and they did so in the name of friendly soccer matches.

The theatrical nature of these relations reached new heights in April 2002 when Koizumi revisited the Yasukuni Shrine. Although Kim Dae-jung issued strong disapproval, he could do nothing more because the World Cup was less than a month away. It became painfully clear just how much officials in both countries would willingly look away from their shared past and present in the name of imagined future national interests.

The lyrics to the official Korea/Japan World Cup song, "Let's Get Together Now," left no room for confusion about the practical understanding of "the unhappy past." Sponsors commissioned young Japanese and Korean pop stars to write and sing a song that a cynic might be tempted to describe as an apology anthem:

Far beyond today, till the end of time
No longer a dream, peace and love becomes reality[41]

Japanese and South Korean politicians and International soccer officials clapped along and tried to dance within the confines of their box seats. The World Cup was a great party, but politicians required a "Don't Worry, Be Happy"-style soundtrack in line with their national interests, yet out of line with lived realities.

"They're only in it for the money"

As Japan and South Korea seek international legitimacy—Japan as a "normal" state and South Korea as "Korea"—the "future-facing" interaction between the two countries has overwhelmed the demands for apology. Meanwhile, the resentment on the part of those unconcerned with the victims' original plight draws strength from the official elision of the past. "They're only in it for the money" is one of the most inhumane attacks lodged against victims of atrocity who are seeking redress. Only privileged members of a society—not necessarily economically privileged, but those privileged by race, gender, or ethnicity—can find a venue to say such things publicly. Advocates of the "in it for the money" rationale use it against people ranging from victims of the atomic bomb and other atrocities to forcibly conscripted soldiers and laborers. The accusation is particularly crass with regard to victims of Japan's organized practice of sexual slavery. In Japan, as elsewhere around the world (such as Bosnia and Afghanistan, to name recent rape camp locations), the "in it for the money" argument relies on two assumptions that entrap victims of sexual slavery and contribute to the difficulty of making their demand for apology heard.[42] First, it presumes that audiences believe to a certain degree that some ethnic groups are less human; women in these groups, therefore, are doubly expendable. Second, it dredges up the "oldest profession in the world" thesis as justification that women—if and when pushed and especially in times of war—will try to profit by selling their bodies. Such assumptions continue to shape historical and political interpretations.[43]

The premises of this claim define the point where the victims' desire for apology runs headfirst into the state's ability to co-opt those desires for national interests. The collision begins when the state utters an apology of some form in response to claims victims have made. Victims, however, remain dissatisfied by what they consider a false apology.[44] Problems compound as victims notice little change in how society treats them, as the state has failed to make the issue of their victimization a matter of social concern. Victims and their advocates often then push the state for a "real" apology, and here the "in it for the money" charge gains momentum.[45]

As mentioned above, the degree to which states participate in apologetic politics depends on their leaders' aspirations to power in the international community. Therefore, if the international community does not demand further apology from the state involved through some sort of multinational pressure, the state's leaders and their supporters assume they have displayed sufficient contrition. Then, when a state's victims counter the state's offer and demand a "real" apology—that is, one they consider meaningful—the victims' critics complain that the state has already apologized and therefore the victims must be "in it for the money." In this respect, the state is made stronger from without and within because it has displayed an amount of remorse accepted by the international formula of normality, yet the state has not fundamentally had to reposition itself. For the time being, state representatives have apologized in the state's national interests. And in the end, although the state's narrators now can discuss the victims' history openly, no one has to take responsibility for this history, let alone the victims' current dignity, or lack thereof. Society again casts off the victims, while their detractors re-empower the old national narrative that existed before the victims' voices were heard.

The cartoonist pundit Kobayashi Yoshinori, the wildly popular self-appointed resuscitator of Japan's "true" past, devotes page after page of his *Sensoron* (On War) series to these ends.[46] His books sell millions of copies, and his growing fame—Japan's leading bookstores prominently feature his books—make his views too prevalent to discount.[47] Kobayashi is pointedly concerned with keeping his own arguments as up-to-date as possible in order to attract young and new audiences. He even delayed publication of *Sensoron* 2 to include the September 11, 2001 attacks on the United States into his explanation of why Japanese should be proud of their nation's historical and "spiritually-based" attempts at "liberating" Asia from white rule.[48]

In *Sensoron* 2, Kobayashi has in effect written a manifesto for the "in it for the money" camp as well as prospective members. He does so by arguing that Japan's conservative establishment sold out the country during the 1990s by even responding to calls for apology from Asian victims of wartime atrocities. One drawing suffices to demonstrate Kobayashi's approach. After a sustained discussion of former prime minister Miyazawa Kiichi's weakness for apology—in one frame Kobayashi has lopped off both Miyazawa's skullcap and that of a foreign ministry spokesman to demonstrate heavy-handedly their "empty-headedness" as they "apologize six times and regret twice!"—Kobayashi depicts a former sex slave as a self-satisfied young woman surrounded by cash flowing from heaven. The text reads:

> Because it was a war-zone and dangerous, the money was great. There were lots of them who earned more than 10 times what a college graduate did in those days and 100 times more than a soldier. In 2–3 years they built houses in their home towns.

On the right-hand side of the drawing, silhouetted soldiers (presumably Japanese) race into explosions while female figures on the other side (presumably like the young woman or her future self) walk away unscathed with bundles and bags—presumably stuffed with money.

Documentary evidence proves that Kobayashi's explanation of Japanese wartime sexual slavery is wrong. The facts are not a matter of opinion at this point. This picture comes in the middle of a diatribe against several of the so-called pragmatic LDP politicians who apologized to Korea in some measure during the 1990s. Although these politicians have never been known for a selfless desire to apologize to victims, Kobayashi's tirade culminates in a demand to "Make 'Insulting the Nation' a Crime and Put (the Politicians) Who Insulted Japan and the Japanese People in Jail!"[49] Ultimately, his growing popularity suggests that Kobayashi's sentiment enjoys wide substantial support.

In Closing

We now have more evidence than ever before of the pre-1945 Japanese government's involvement in atrocities such as sexual slavery and forced labor. At the same time, the state's protagonists have never been in a stronger position to transform such material into internationally acceptable apologies for the nation's interests at the expense of the dignity its historical victims seek.

In the hands of the state, the current pragmatic approach toward apologizing for the past through the formula of "sorrow and regret" threatens to steal history away from the individuals who lived through the horrors. When victims find fault with the state's formula, prideful zealots denounce those who speak out on the victims' behalf as espousing a "masochistic" view of Japanese history, a charge that Japan's leaders are either failing to challenge or are encouraging themselves.

For victims, apology is something with a clear use value to them: it acknowledges their suffering and grants them dignity. One former sex slave has recently defined how such dignity could be gained: "The emperor should bow at my feet."[50] To such victims, the issue is not what counts as a pragmatic approach. What counts is what is meaningful to them. But because the international formula does not require that the state allow individuals to define apology for themselves, while the surviving victims dwindle in number, Japan's leaders knowingly play the "future-facing" waiting game.

Notes

1. See Elazar Barkan, *The Guilt of Nations: Restitution and Negotiating Historical Injustices* (New York: Norton, 2000).

2. For the North Korean example, see *Donga Ilbo* and *Asahi Shinbun*, July 26, 2002; August 3–4, 2002; *New York Times*, web edition, July 25, 2002. For the IRA, see *An Phoblacht* as well as the *New York Times*, July 16–17, 2002; see also BBC International coverage, July 16–18, 2002. Tellingly, in December 2002, Saddam Hussein offered an "apology" to Kuwait, but the Kuwaiti government refused to accept it and encouraged him to take his case to the UN. See http://news.bbc.co.uk/2/hi/middle_east/2555515.stm

3. See Ian Buruma, *The Wages of Guilt: Memories of War in Germany and Japan* (New York: Farrar, Straus & Giroux, 1994).

4. Left-leaning booksellers such as Tokyo's Ajia Bunko and Mosakusha were the first to create separate sections, but mainstream stores such as Horindo and Kinokuniya have followed suit. Under Arai Shinichi's direction, the Center for Research and Documentation on Japan's War Responsibility publishes the quarterly journal *Senso Sekinin Kenkyu* (The Report on Japan's War Responsibility) and in early 2002 Iwanami publishers reissued Ienaga Saburo's 1985 benchmark analysis, *Senso Sekinin* (War Responsibility).

5. The Research Institute for Peace and Security in Tokyo, e.g., pointed to "the history problem" as a major sticking point in regional cooperation. See *Heiwa/Anzen Hosho Kenkyujo* (The Research Institute for Peace and Security), ed., *Ajia no Anzen Hosho 2001–2002* (Asian Security Policy) (Tokyo: Asagumo Shinbunsha, 2001), 37–39, 212–213.

6. Arai Shinichi and Iko Toshiya compiled a compendium of claims in *Sekai* (World), No. 696 (December 2001), 178–196. Utsumi Aiko discusses this trend in her conclusion to the 1999 reissue of *Sengo Hosho to wa Nanika* (What is Postwar Compensation?) (Tokyo: Asahi Bunko, 1999 [1994]), 195–205. See also T. Fujitani, Geoffrey M. White, and Lisa Yoneyama, eds., *Perilous Memories: The Asia Pacific War(s)* (Durham: Duke University Press, 2001).

7. See, e.g., Kobayashi Yoshinori and Nishibe Susumu, *Hanbei to iu Saho* (A Handbook for Anti-Americanism, or, Anti-American Etiquette) (Tokyo: Shogakkan, 2003).

8. For an insightful elaboration, see Jae-Jung Suh, "The Two-Wars Doctrine and the Regional Arms Race: Contradictions in US Post-Cold War Security Policy in Northeast Asia," in *Critical Asian Studies*, 35: 1 (March 2003), 3–32.

9. Security Council Resolution 1325 (October 31, 2000) requires member states to condemn the practice of targeting women during war, and, in particular, the practice of sexual slavery. Should Japan gain the permanent seat it desires, it will have to sign this resolution, which could strengthen the claims against Japan for redress for Japan's notorious "comfort women" system.

10. For a well-written version of this approach, see Funabashi Yoichi, "Kako Kokufuku Seisaku o Teisho Suru" (A Proposal for a Policy to Overcome the Past) *Sekai* (World), no. 692 (2001): 48–62, especially 50–51. See also Victor D. Cha, *Alignment Despite Antagonism: The US–Korea–Japan Security Triangle* (Stanford: Stanford University Press, 1999); and Michael J. Green, *Japan's Reluctant Realism: Foreign Policy Changes in an Era of Uncertain Power* (New York: Palgrave, 2001).

11. President Bill Clinton ran into trouble when he "apologized" for slavery on his 1998 Africa trip, and his advisers subsequently went to great lengths to make sure he only "regretted" other horrors. When the U.S. press "discovered" the No Gun Ri massacre in 1999, Clinton's "regretful" response was in line with this policy, and the U.S. army report

likewise officially described No Gun Ri as a "regrettable accompaniment to war." Quoted in Charles J. Hanley, Sang-Hun Choe, and Martha Mendoza, *The Bridge at No Gun Ri: A Hidden Nightmare from the Korean War* (New York: Henry Holt, 2001), x. Official materials concerning the matter remain classified, but in off-the-record discussions, two U.S. intelligence officers confirmed to me a concerted effort to avoid the word "apologize." Reports in *The New York Times* substantiate this. See December 22, 2000 and January 12, 2001.

12. Norma Field, "War and Apology: Japan, Asia, the Fiftieth, and After," *Positions: East Asia Cultures Critique; Special Issue: The Comfort Women: Colonialism, War, and Sex*, 5: 1 (1997), 5.

13. Only the rarest of Japanese politicians has publicly noticed that the rhetoric of "sorrow and regret" itself sustains the problem. On the eve of South Korean president Roh Moo-hyun's inauguration, Japanese parliamentarian Okazaki Tomoko visited Seoul to speak with him. Later she informed reporters that she "requested (President-elect) Roh to take the issue (of sexual slavery) seriously. As Roh emphasizes a 'future-oriented' relationship, I pointed out that the future can only begin after clearing up the past." Quoted in *The Korea Herald*, February 13, 2003.

14. For documents concerning Japan–South Korea normalization, see United Nations, *Treaty Series: Treaties and International Agreements Registered or Filed and Recorded with the Secretariat of the United Nations*, 583 (New York: United Nations Publications, 1966). For a compelling discussion of Japan's formula for redress in Southeast Asia, see Sayuri Shimizu, *Creating a People of Plenty: The United States and Japan's Economic Alternatives, 1950–1960* (Kent, OH: The Kent State University Press, 2001), especially ch. 4.

15. September 6, 1984, reception at the Imperial Palace, Tokyo, quoted in Arai and Iko, 188.

16. Karen L. Gatz, ed., "US Efforts to Encourage Normalization of Relations Between the Republic of Korea and Japan," *Foreign Relations of the United States, 1964–1968: Volume XXIX, Part 1, Korea* (Washington, D.C.: U.S. Government Printing Office, 2000), 745–802.

17. In Gatz, No. 342, Komer to Bundy, May 19, 1964, 760 (italics in original). With U.S. expenditures in Vietnam escalating, this view rapidly gained momentum.

18. In Gatz, No. 349, Reischauer to Rusk, September 8, 1964, 770.

19. In Gatz, No. 353, Reischauer to Rusk, November 21, 1964, 778.

20. *Asahi Shinbun*, February 17, 1965, front page, evening edition.

21. *Asahi*, February 17, ibid.

22. In Gatz, No. 357, Thomson to McGeorge Bundy, February 20, 1965, 784.

23. Okuda Masanori, "Sengo Hosho Saiban no Doko to Rippoteki Kaiketsu" (Movements in Postwar Compensation Decisions and Their Legal Interpretations) in Chi Myon Gwan, *Nikkan no Sogo Rikai to Sengo Hosho* (New Japan–Korea Partnership and Postwar Compensation) (Tokyo: Nihon Hyoronsha, 2002), 131–146 and especially the useful chronological chart of claims against Japan from 1990 to the present, 147–159.

24. William Slany, State Department Historian, in Gatz, ix.

25. E-mail communication with the author, June 12, 2002.

26. Interview with Imazu Hiroshi, Tokyo, June 11, 2002.

27. Imazu, "Souru no Yokkakan" (Four Days in Seoul) *Asahi Shinbun*, February 21, 1965, 12. For more discussion, see Imazu, *Jyanarisuto Sono Yasashisa to Tsuyosa: Kingendaishi e no Arata na Tabidachi* (A Journalist's Strength and Kindness: A New Journey into Modern History) (Tokyo: 3A Network, 1998), 186–196; also, Imazu, "Kimpo Kuko ni Kieta Kimigayo: Shiina Gaisho, Hokan no Asa" (Canceling 'Kimigayo' at Kimpo Airport: The Morning of Foreign Minister Shiina's Arrival in South Korea) in *Nihon Kissha Kurabu Kaiho* (Reports from the Japanese Journalists' Club), no. 377 (2001), 8–9.

28. *Asahi Shinbun*, February 21, 1965. Imazu noted that many Koreans distrusted Japanese intentions because the national income of Korea was one-tenth of Japan's, and the technical and industrial levels were so far apart.

29. During the preparation for most recent normalization talks in August 2002, North Korea's demands for "apology" and "compensation" emerged as a chief obstacle. See *Asahi Shinbun*, August 26, 2002.

30. See coverage in *Asahi Shinbun*, October 1998; *Chosun Ilbo*, October 1998; *New York Times*, October 1998. The full text of the October 1998 partnership declaration is in Arai and Iko, 195.

31. Michael Robinson has written extensively on Japan's initial era of "cultural sharing" with Korea during colonial rule. See Robinson, "Broadcasting, Cultural Hegemony, and Colonial Modernity in Korea, 1924–1945," in Gi-Wook Shin and Michael Robinson, eds., *Colonial Modernity in Korea* (Cambridge: Harvard University Press, 1999), 52–69.

32. See *Hankyoreh and Asahi Shinbun*, June 7, 2003.

33. Kim Dae-jung and Okamoto Atsushi, "Kokuminteki Koryu to Yuko no Jidai o," (Towards a More Citizen-Oriented Pattern of Exchange) in *Sekai*, No. 653 (1998), 61.

34. Mizuno Takaaki, " 'W' Hai ni Miru Nikkan no Kyokan to Nashyonarizumu" (Viewing Commonalities and Nationalisms Between Japan and Korea Through World Cup), paper presented at the University of Tokyo's Minority Studies Workshop, July 16, 2002.

35. For discussion of how the textbooks have played recently into Korean–Japan relations, see Yi On Sun, Chon Ji Jeon, and Ishiwata Nobuo, *Kankokuhatsu: Nihon no Rekishi Kyokasho e o Hihan to Teigen* (The View from Korea: Objections and Criticisms to Japanese History Textbooks) (Tokyo: Kirishobo, 2001).

36. See Hata Nagami, "Koizumi Shusho Yasukuni Sanpai no Seiji Katei: 'Kokka to Ireisai' ni Kansuru Joron," in *Senso Sekinin Kenkyu*, 36 (2002), 10–18. Shortly after Koizumi entered office in the spring of 2001, Ian Buruma made almost the opposite point about Japan's official disregard of Asian claims. (Buruma, "The Japanese Berlusconi?," *New York Review of Books*, July 19, 2001, 42–44.) Buruma maintained that under Koizumi, "We might get a great deal of hot air about the superiority of Japanese culture, the justice of Japan's war, the racist arrogance of the United States, the need for a strong Japanese defense state and so on, but if in the end the result is a more open, more democratic Japan, that might just, but only just, be worth putting up with" (44). Koizumi's arrogance did not play out well around the world until attention was diverted by the U.S. war first in Afghanistan, then Iraq, whereupon it was forgotten; a "more democratic Japan" has not emerged.

37. This myopia plays out not only with regard to Asian victims making claims against Japan. Shortly after the new war began, President George W. Bush asked Congress to drop a provision in a spending bill that encouraged former U.S. prisoners of war to sue Japanese companies for World War II forced labor. Congress removed the provision to avoid undermining relations with Japan which it wanted squarely on board in its war preparations. Kyodo News Service Online, November 30, 2001. In January 2003, on the eve of war in Iraq, a U.S. appeals court in San Francisco barred veterans' claims as "unconstitutional." *The New York Times*, online, January 21, 2003.

38. Prime Minister Koizumi followed the practice of one-day apology tours in his historic September 17, 2002 visit to Pyongyang to open normalization negotiations between Japan and North Korea. This trip took an unplanned turn, however, when North Korean leader Kim Jong Il apologized for his country's past abductions of Japanese citizens. Koizumi signed the provisional agreement in which Japan "regrets the past," but the ramifications of the "double apology" remain uncertain. *Asahi Shinbun*, September 17, 2002, evening edition.

39. For the China visit, see *Asahi Shinbun*, October 9–11, 2001 (morning and evening editions); for Korea, see October 15–16, 2001 (morning and evening). Not surprisingly, in Seoul Koizumi made a linguistic miscalculation, tripping over his statement of regret to encourage Japanese and Koreans to "reflect on the past together"—the "together" element sparked further protests in Korea. See *Asahi*, October 18, 2001 as well as related articles in the online edition of the *Chosun Ilbo* (Japanese edition), October 17 and 22, 2001. My thanks to *Mainichi Shinbun* editor Shimokawa Masaharu for calling my attention to the latter.

40. *Asahi Shinbun*, March 23, 2002.

41. 2002 FIFA World Cup Official "Korea/Japan" Song, "Let's Get Together Now," Lyrics by Sawamoto Yoshimitsu, Matsuo Kiyoshi, Lena Park, and Kim Hyung Suk, Copyright DefSTAR Records, 2002. Because a ban on Japanese pop culture was still in effect, the Korean government had to grant special dispensation for the sale of this CD, due to the Japanese lyrics. See *Asahi Shinbun*, March 23, 2002.

42. See Women's Caucus for Gender Justice, *Public Hearing on Crimes Against Women in Recent Wars and Conflicts: A Compilation of Testimonies* (New York: Women's Caucus on Gender Justice, 2000).

43. See Yuki Tanaka, *Japan's Comfort Women: Sexual Slavery and Prostitution During WWII and the US Occupation* (London: Routledge, 2002).

44. This point is poignantly revealed in Byun Young-Joo's 1997 film, *Habitual Sadness: Korean Comfort Women Today* (International distribution: Filmakers Library, New York); see Chungmoo Choi, "The Politics of War Memories toward Healing," in T. Fujitani, Geoffrey M. White, and Lisa Yoneyama, *Perilous Memories*, 395–409.

45. A well-respected former professor of economics at the University of Tokyo, Nishibe Susumu, recently joined ranks with those decrying victims' demands for compensation and apology from Japan. Nishibe added his veneer of sensibility to the routine blaring of sound-trucks in Tokyo and Osaka and throughout Japan, and argued that Japanese "should learn from Chinese and Koreans" about getting apologies (in particular, Nishibe said Japanese should follow the tactics of the "whiny Koreans") to get the apology Nishibe argues the United States owes Japan. See Kobayashi and Nishibe, *Hanbei to iu Saho*, 96.

46. Kobayashi Yoshinori, *Sensoron* (On War) (Tokyo: Gentosha, 1998); *Sensoron 2* (Tokyo: Gentosha, 2001).

47. Recent academically inclined versions of Kobayashi-like vitriol with regard to Korea include Nakagawa Yatsuhiro, *Rekishi o Gizo Suru Kankoku: Kankoku Heigo to Sakushu Sareta Nihon* (How Korea Distorts History: Korean Annexations and the Exploitation of Japan) (Tokyo: Tokuma Shoten, 2002) (217–248 for discussion of sexual slavery); and Sato Katsumi, *Nihon Gaiko wa Naze Chosen Hanto ni Yowai no ka* (Why Is Japanese Foreign Policy So Weak Towards the Korean Peninsula?) (Tokyo: Soshisha, 2002). In their introduction to a critique of Kobayashi's first *Sensoron*, editors wrote that to ignore Kobayashi's popularity was not academic snobbery, it was "cowardice." Umeno Masanobu and Sawada Tatsuo, eds., *Sensoron/Bosoron* (On War/On Irresponsible Thinking) (Tokyo: Kyoiku Shiryo Shuppankai, 1999).

48. Kobayashi, *Sensoron 2*, 9–31

49. Ibid., 298.

50. Author's notes from witness testimony given at the Women's International War Crimes Tribunal on Japan's Military Sexual Slavery, Tokyo, December 2000.

Suggestions for Further Reading

Field, Norma, "War and Apology: Japan, Asia, The Fiftieth, and After," *Positions: East Asia Culture Critique*, 3: 1 (1997): 1–49.

Fujitani, Takashi, Geoffrey M. White, and Lisa Yoneyama, eds., *Perilous Memories: The Asia-Pacific War(s)* (Durham, NC: Duke University Press, 2001).

Tanaka, Toshiyuki, *Hidden Horrors: Japanese War Crimes in World War II* (Boulder, CO: Westview, 1996).

Yoshimi, Yoshiaki, *Comfort Women: Sexual Slavery in the Japanese Military During World War II*, translated by Suzanne O'Brien (New York: Columbia University Press, 2002).

Yoneyama, Lisa. "Traveling Memories, Contagious Justice: Americanization of Japanese War Crimes at the End of the Post–Cold War." *Journal of Asian American Studies*, 6: 1 (2003), 57–93.

CHAPTER 3

Breaking the Silence in Post-Authoritarian Chile

Katherine Hite

From the 1980s through the close of the twentieth century, countries throughout Latin America, from the southern tip of South America to Central America and Mexico, experienced a major wave of democratization. In contrast to the 1970s, when democracies in the region were few and far between, by the end of the century all but one Latin American country, Cuba, was a formal democracy.

One of the central issues Latin American democratizing and re-democratizing regimes have had to face is how to address, or "come to terms with," what in many cases were massive, systematic violations of human rights under the military dictatorships they replaced. In Guatemala, an estimated 100,000 citizens were killed in army counter-insurgency campaigns and death squad attacks during a military regime that lasted over 30 years (1954–1986). Under the Brazilian military dictatorship (1964–1985), thousands of suspected subversives suffered systematic torture and imprisonment. At several points during its 11-year reign, the Uruguayan military regime (1973–1985) possessed the highest political prisoner population per capita in the world.

Democratizing regimes have approached such traumatic, violent pasts in a range of ways. Argentina, for example, became the first and only democratizing regime to put all nine of the country's former military junta leaders on trial for the murders and disappearances of scores of the country's citizens during the 1976–1983 dictatorship. Other democratizing governments established official "truth commissions" to investigate and report on their painful pasts, and several adopted compensation policies for the families of human rights victims. However, policies to address the abuses of the past have been both fitful and highly contested. Latin American states and societies continue to be enmeshed in soul-searching debates about what dimensions of their painful histories should be remembered and who should be held responsible.

Chile was among the last of the Latin American countries to undergo a transition from military rule (1990). On one level, this seemed paradoxical. Prior to Chile's military coup d'état on September 11, 1973 (a fateful anniversary the United States now shares), the country boasted a longstanding multiparty democracy virtually unparalleled in the region. Even Chileans who supported the 1973 coup never anticipated that the suspension of democracy would last so long nor prove so violent. Nevertheless, Chilean General Augusto Pinochet remained in power for 17 years, and his dictatorship irrevocably altered Chilean politics and society.

The brutality and duration of Chile's Pinochet regime instilled deep-seated and lasting fears, diminishing the individual's sense of security as well as individual desires to participate in political life. The fact that Chileans had not known such repression prior to the dictatorship exacerbated the long-term traumatic effects of a regime that disappeared and executed more than 3,000 citizens, while an estimated one in every ten families experienced arrest, torture, and/or exile. Targeted, direct repression had a multiplier effect: kidnapping, torturing, and disappearing one person affected the many.

The violence wrought by the military and the police continue to shape the political imagination of youth today. In an informal interview with third-year students in the history department of the University of Chile, students discussed why they do not participate in the protests that have occurred over the last few years at the university. Daniela responded, "Fear is very strong. The police show their strength whenever there is a march or a protest and many people do not come out to participate because they are afraid." Another student, Victor, added, "So many students from this campus died during the coup and the dictatorship, and we are here every day and that memory has not faded."[1]

In addition to traumatic memories of the brutality of the dictatorship, Chileans possess deeply divided memories and interpretations regarding what led to the military regime in the first place. During the preceding years, the country polarized politically along class and ideological lines. Polarization grew acute under Chilean socialist leader Salvador Allende, who in 1970 became the world's first democratically elected Marxist president. Pinochet's coup ended Allende's Popular Unity (UP) administration three years later, and Allende died amidst the air force bombardment of the presidential palace. Pinochet sought to eliminate the Chilean left and the organized social bases of leftist party support.

Memories of Allende's 1970–1973 "Chilean road to socialism" are fraught with images of long lines for bread and cooking oil, massive mobilizations and counter-mobilizations, conflict and tension in the workplace, schools, neighborhoods, and families, both for those who sympathized with the UP government and those who did not. The Pinochet dictatorship consistently repeated the message that life under the Allende government was chaotic, even dangerous, that the military had saved Chile from communism and from potential conflagration, even civil war, and that a return to democracy would have to be controlled and protected. Chileans today both honor and vilify Allende. He is remembered as a victim of the military, the Right and the United States, a victim of his own party and the ultra-left, a martyr, a disastrous leader, a sinister protagonist seeking to unravel Chilean society.

During the 1980s struggle for a return to democracy, two historic political enemies, Allende's Socialist Party and the Christian Democratic Party, forged an alliance against the dictatorship that would eventually carry them into government. Since 1990, there have been three successive, democratically elected, multiparty *Concertación* administrations, two of which were headed by Christian Democratic presidents (Patricio Aylwin from 1990 to 1994 and Eduardo Frei, Jr., from 1994 to 2000) and the current by a Socialist (Ricardo Lagos 2000–2006). Consensus politics has been the norm in the Chilean executive and legislature, making Chile a model of democratic governance in the region. Yet not far beneath the surface of this consensus are tensions rooted in memories of the full-blown conflicts of the 1970s and their sobering results.

While a good deal of time has passed since the 1973 coup, only recently have Chileans truly begun to engage in a public—albeit divided—remembrance of the overthrow of Allende and the repressive years of the dictatorship. The unanticipated October 16, 1998, arrest of Pinochet in London most forcefully contributed to what had become a steadily increasing series of explorations unearthing the horrors of the dictatorship and those who defended it, as well as the intense political conflicts of the preceding democratic era.

In a recent survey by the Chilean group Fundación Futuro, the vast majority (87 percent) of Chileans agreed with the statement: "in order to overcome the hatreds of the past it is necessary that the truth be known and justice be done."[2] Such attitudes reach beyond desires for a mere exploration of recent history. Yet the task of historicizing Chile's recent past is circumscribed by competing political agendas as well as all too recent traumas that exacerbate the challenges and the politicization of the task. Moreover, Chilean political leaders have been the last to engage in public remembrances of the past.

Silences

During the course of Chile's gradual re-democratization process, analysts have employed the notion of a "pact of silence" among the political elite regarding the past. For politicians of today's governing left, several of whom served in the Allende government more than two decades before, the pact refers to the Left's reticence to challenge rightist, Pinochet-endorsed narratives of the 1973 breakdown of the democratic regime and the repression that followed. Silence on the Right refers to a refusal to assess rightist contributions to the 1973 breakdown, as well as reticence to question the tactics or policies of the dictatorship.

Indeed, such silence transpired through most of the 1990s in spite of the increasingly abundant range of investigative reports, testimonies, documentaries, films, memoirs, and other accounts of the last 40 years. Silence prevailed among the governing elite even when, in the aftermath of Pinochet's 1998 London arrest, dozens of former officers and civilians began to be prosecuted for human rights violations, and daily reports began to appear in even the most conservative press regarding past human rights violations and violators. While the governing Left consistently refused to accept centrist and rightist proposals for a statute of limitations on human rights violations charges, Left leaders also largely resisted proactive stances on coming to terms with the past. In fact, explorations of Chile's past have largely occurred in spite of the political elite, rather than because of it.

More often than not, analysts attribute the elite-level silences to what might generally be termed rational calculations based on Chile's political, institutional design and the correlations of power represented by it. During the transition process, the opposition to the dictatorship was unable to break the might and autonomy of the military and its authoritarian and rightist supporters. Unlike the discredited Argentine military that in addition to a record of systematic human rights violations left in its wake a disastrous economy and a defeat by the British, the Chilean military could take credit for a comparatively healthy economy, and remained cohesive and powerful. The Chilean armed forces would fundamentally determine the transition process and retain important institutional guarantees that effectively grant veto power against attempts to confront authoritarian enclaves embedded in the 1980 military-sponsored Chilean constitution and in other spheres of social and political life.[3]

In comparison to most democracies, the Chilean military enjoys unusual autonomy from civilian authority over key dimensions of the military as institution. For example, the president does not appoint the commanders-in-chief of the military branches. Moreover, the military-designed electoral system for determining legislative seats deliberately overrepresents the political minority, namely the Chilean right. In addition, the Pinochet regime placed nine military-appointed senators-for-life in the Congress, and Pinochet wrote himself into post-authoritarian politics as a senator-for-life upon his retirement as commander-in-chief of the army. Given the political alignment of forces, many argue that attempts to resurrect a past that places authoritarian incumbents in a bad light only threatens to bring gridlock to a system that requires such incumbents' consent for even modest reforms.

Examinations of the past also tend to unmask the historic enmity between the political center and Left, the two forces that today compose the governing alliance. Socialist Party senator and former cabinet minister Carlos Ominami holds that reluctance to examine the past is due to strategic concerns that threaten the Socialist Party–Christian Democratic alliance:

> . . . There is no question that there is a real complicity—objective complicity, not subjective—on everyone's part. For the right, it's obviously due to cowardice, for it doesn't benefit them, but there is something of a pact of silence . . . So much isn't known because the political right isn't served by anyone knowing, and because sectors of the left and the Christian Democrats aren't served, either. The Christian Democrats look up at the ceiling when these themes surface . . . When Aylwin became president [from 1990–94], I was one of his ministers, but Aylwin strongly supported the coup! . . . So a big fat theme is that it doesn't help too much to remind ourselves of all this . . .[4]

In a slightly different vein, analysts claim that the silences regarding the past are due to what might be termed "political learning." The Left's major defeat in 1973 engendered a profound series of reflections on the Popular Unity government's failures. According to political learning arguments, such reflections contributed to Left sectors' new "valuation" or appreciation of the democratic process, including the spirit of compromise and the need for majoritarian coalitions.[5] According to socialist senator Ricardo Nuñez, central to the socialists' political learning process was also the crucial need to prove their ability to govern:

> . . . We had to show we could govern well. And the countries that govern well are those that recognize the socioeconomic realities of their countries. And in the first

years the economic success of the transition was spectacular. Because we had a budget surplus, 7% growth rates, we moved one million poor people out of absolute poverty, the market flourished, and we felt this had to be a constitutive factor of the Chilean transition. Political success, economic success. This required not returning to the past, not returning to '73.[6]

Such "lessons learned" have discouraged governing leftist leaders from resuscitating memories of conflict and memories of failure. Yet elite-level silences have hampered society's efforts to come to terms with the past. Leaders are central to creating the norms, rules, and institutions that govern or frame societal explorations, and coming to terms with historical legacies are in good part contingent upon the articulation of such legacies between elite and citizenry. In addition, in the context of the consistent uncovering of the traumas of the past, elite-level silences arguably affirm, even exacerbate, widespread societal disaffection with Chilean politics.

The Many Layers of Trauma

For those who were part of the 1970–1973 UP leadership, memories from one traumatic realm predominate—the UP's dwindling control while in power and the absolute loss of control thereafter. These memories trigger a profound sense of guilt for the infighting and mistakes of Allende's government, particularly regarding the economy and private property policies. While the UP platform advocated a Chilean road to socialism, there was a great deal of disagreement within the UP regarding the means to achieving such ends, particularly given that Allende had won the presidential election with only 36.7 percent of the vote. Arguing the need to win over larger segments of the population for radical social change, leaders of the Chilean Communist Party tended to urge a more gradual, cautious pace of reform. In contrast, the Socialist Party leadership favored more rapid reform, including the immediate nationalization of major foreign-owned firms and the expropriation of leading private sector industries and agricultural holdings. The socialists argued that a failure to carry out the rapid transfer of Chile's leading industries into state hands would only allow the Chilean bourgeoisie the opportunity to engage in both major capital flight and in sabotage. These and other intra-UP disputes made policymaking difficult and unpredictable, as the UP became increasingly vulnerable to external opposition.

Former UP government officials describe their memories of anxiety and dread, most powerfully felt in 1972, during the strikes, sabotage, and other forms of organized opposition to the UP, sponsored both from within the country and from outside, principally by the United States. While the UP picked up greater electoral support in the 1971 and 1973 congressional elections, the election results epitomized society's polarization, as the coalition obtained close to 50 percent of the electorate, while the opposition parties captured the other half. Unable to obtain enough electoral support to constitute a firm majority in the Congress, several opposition sectors threw their weight to an "extra-constitutional" way out—a military coup d'état. For members of the UP coalition, the defeat of September 1973 confronted UP leaders and militants with the deaths of comrades, massive arrests, torture, exile, the breakup of collectivities and families, and a great deal of fear.

Such traumatic defeat contributes to a continuing sense of powerlessness among a significant sector of the current leftist leadership, even though the Left holds the political reins. There is a paradoxical relationship at play: leftist leaders invariably perceive the traumas they and others experienced in the past to be a result of their own doing, and yet this sense of responsibility paralyzes them before their contemporary opponents. The traumas become internalized as ever-present dimensions of individuals' political identities and choices.

The Trauma of Victory

The experience of political victory, namely the 1970 victory of the UP government, was as crucial a traumatic experience as that of subsequent defeat. According to UP education minister and Socialist Party leader Aníbal Palma, being in power was the most profound and difficult experience in his life.

> Before the UP government, I had never held a government position. In September of 1972 Allende appointed me undersecretary of foreign affairs, and shortly thereafter minister of education in an extremely conflictive period . . . It was extremely hard . . . and I had been both a high school and a university student leader, so when I had to confront student conflicts, and I saw them marching in the streets, screaming slogans against the government exactly as I had done before, I felt as if I were living a dual personality. I remember several times receiving dispatches to go and see student demonstrations that attacked the ministry, and I had done exactly the same thing . . . We were living the world in reverse!
>   . . . I tell you all this frankly, I believe that I have never lived more bitter moments in my entire life than in the moment that the opposition students took to the streets, and there were fights, making it necessary for the police to intervene. I always lived with the fear that at some point a student would be killed or terribly injured, and I felt responsible for whatever might happen, and each protest gave me an enormous sense of tension, and it felt so out of my hands . . .[7]

Palma's sense of role reversal in power, his sense that he had lost control of the moment, echoes in the accounts of other former UP leaders and militants. Such memories arguably discourage the current Socialist Party leadership from advocating participatory politics that include public mobilization, or even active party recruitment, contributing to what analyst Alexander Wilde termed the "muffled" quality of 1990s Chilean public life.[8]

For the current vice-president of the Chilean House of Representatives, Adriana Muñoz, memories of her role as a UP government official and a Socialist Party militant highlight the intensity of her absorption in a charged political moment. Muñoz became an undersecretary in the Allende administration's Department of Agriculture. She was 22 years old. Muñoz recalls being "swallowed up" in the intensity of party militancy during a period of "a lot of rage," and of being "oblivious to the risks." Such memories have served Muñoz as signifiers for what to avoid politically. Today Muñoz struggles within the Congress for social reforms—most notably a divorce law—but does not advocate civil society mobilizing on the law's behalf. Nevertheless,

Muñoz laments the silences:

> The country is afraid of debate. Or we as the beaten-down left are afraid to be
> considered as adding once again to divisiveness, destruction. There is a very strong
> trauma within us, and perhaps generations of the future will be able to recuperate
> these memories with greater lucidity . . . We are afraid. We are extremely sensitive
> to the fragility of the process, and the fragility of our power.[9]

The Trauma of Resistance

For those on the Left who were armed combatants at particular stages of struggle
against the dictatorship, there is another and distinct temporal realm of trauma—of
comrades dying, of physical agony, and for some, a questioning and profound
remorse for subjecting fellow comrades to what proved a futile effort. For former
Movimiento de Izquierda Revolucionaria (MIR) political organizer Patricio Rivas, now
an Education Ministry official, dead comrades are the central referent of his life account.

Unlike many of his comrades, Rivas survived his incarceration and torture and
spent the better part of a decade in exile. In 1984 he returned clandestinely to Chile,
still committed to the MIR, which for the past several years had pursued a guerrilla
strategy that proved suicidal against the dictatorship. Rivas could not escape the
tremendous sense of responsibility he felt for being a part of such a fatal strategy.
Rivas recounts the arguments and division among his comrades over whether to risk
lives in the name of resistance and defeat of the dictatorship.

> We . . . concluded that there was no possible way to correct the great error we had
> made [of returning militants to fight the dictatorship] . . . We fought, we fought,
> we fought so much, and I would arrive home and I would think, "But we're fighting
> and we're how many, 300, 500?" It was a difficult, difficult period.[10]

Rivas's life account blends several layers of pain and trauma—the experiences of party
division, comrades dying, collective responsibility, intense physical agony from past
torture, and the irony of feeling like a traitor because at one point he had to leave the
country (if only momentarily) for back surgery. Yet in spite of the pain of feeling
responsible for the deaths of comrades, of internal party splits, and accusations of
betrayal, Rivas expresses an enormous pride in having been a part of the MIR and in
the resistance, a pride seldom heard in contemporary Chilean analyses of the past.

> One does what one knows. And that's what I knew, and I'm very proud of what
> I did. Very proud . . . And I continue to believe that you, me, all of us are the last
> monkeys, but the first man has yet to appear. The project of a man who is pro-
> foundly humane has not yet appeared. And this world will end with us if we are
> not capable of producing certain changes . . .
> Many times I have felt guilt for being alive. Because I am a statistical error.
> But I don't know why, it's chance, so I have this responsibility to history and to
> my history to continue thinking with the same courage.[11]

Rivas's memories and sentiments represent a sector of Chilean society that has played a very vocal, testimonial role in pressing for an accounting of the past. In contrast to trauma that has a silencing effect, the trauma of memories of personal and collective sacrifice against the dictatorship arguably fuel a continued public resistance on the part of the survivors—those who consider themselves "statistical errors"—directed against social injustices and any impunity for the authoritarian regime. It consistently places the nongoverning Left in angry stances against the government.

Resurrecting Trauma

While it is generally claimed that silence has prevailed among members of the Chilean governing elite from the very outset of the Chilean transition, a closer read reveals that in 1990, the first year of the transitional government, congressional leaders from Left to Right were far more outspoken regarding the past than in any subsequent year. The executive's emphasis on "truth" about the recent past paved the discursive arena, while the discovery of previously disappeared, mummified bodies in northern Chile heightened the ethical–symbolic tone of the debate. The year 1990 represented a space of official discursive debate about the last 40 years, albeit bounded by consistent hints that truth-telling would not be accompanied by a demand for justice. However, elite debate was comparatively short-lived. Right-wing Unión Democrática Independiente (UDI) senator Jaime Guzmán's assassination on April 1, 1991, traumatized the political elite, and official discourse on the past shut down.

President Patricio Aylwin set the leadership agenda by placing the need "to clarify the truth and do justice to the subject of human rights, as a moral exigency unavoidable for national reconciliation" at the top of his five-point program of governmental tasks. The Aylwin government established a blue-ribbon truth commission charged with responsibility for investigating the details of deaths and disappearances (though not those who were tortured and survived) under the dictatorship. Working behind closed doors for six months, the Truth and Reconciliation Commission ultimately produced a synthetic, eloquent interpretation of the country's past quarter-century, from the Allende years to 1990. The commission provided specific documentation of those who had been executed and disappeared, and it made a series of recommendations regarding compensation for victims' families. Citing the need to leave judgment of guilt or innocence to the courts, the commission did not publicly name known human rights violators.

During the first several months of the new government, opposing interpretations of history echoed in virtually every issue up for congressional debate. Socialist Party senator Ricardo Nuñez repeatedly called for the scheduling of a formal Senate session on the 1970–1973 years. Rightist senators, such as Renovación Nacional (RN) senator Ignacio Pérez Walker, and Pinochet-appointed retired generals Santiago Sinclair and Bruno Siebert, devoted much of their time on the Senate floor presenting their interpretations of the history of the military regime. In these early months of the *Concertación* administration, no one account of history publicly prevailed over another.

The Christian Democratic discourse on the Senate floor was often scriptural in tone, emphasizing the need to document the recent past in order to heal a "wounded Chile," in order to forgive, to "exercise man's two great potentials: his capacity to know and search for truth—the motor of history—and his capacity to love that brings him closer to God."[12] Socialist Party discourse generally proved more condemning of the dictatorship. Senator Nuñez focused on the "Pinochet regime's systematic violation of human rights," emphasizing the wide range of violations, and the need to specify the facts. Nuñez personalized the victimization:

> Those who, for example, were incarcerated for long months, and became a victim of exile for the sole fact of having been vice-rector of the second most important university in the country, have a great attachment to the truth and to justice, but never to revenge, nor much less to violence.[13]

Right-wing politicians demanded the truth commission reframe its understanding of violent history, first by expanding the commission's investigative timeline to include the 1960s. UDI senator Eugenio Cantuarias insisted that the commission recognize that "the violence in Chile originated in the second half of the 1960s":

> . . . in order to paint a complete picture [we must] own up to what was the progressive deterioration of national coexistence, of respect for individuals' essential rights, especially the right to property, which originated with the Agrarian Reform Law [of the 1964–1970 Christian Democratic Eduardo Frei administration] and continued with its systematic application. It implied that a number of Chileans came to be second-class citizens, given that their most elemental rights could be so vulnerable, that they had no recourse to defend themselves. This meant that sectors of the national citizenry who were traditional defenders of order began to consider the use of force legitimate as well, as a last recourse to defend their earned rights. Thus, then, the action of the radicalized Left and the populist demagoguery to which those governing [in the 1960s] had fallen prey are the indispensable antecedents that must be on the table.[14]

Cantuarias's narrative resuscitates what for the Chilean ruling elite proved perhaps the most traumatic moment in contemporary history—the expropriation of private property in the Chilean countryside. In order to ensure that the then-presidential candidate Salvador Allende lost the 1964 elections, the Right had supported Eduardo Frei Montalva's candidacy. The Chilean landed elite and their rightist political allies viewed the Agrarian Reform Law of the Frei administration as an unacceptable betrayal. More profoundly, the reform marked the most visible sign of a breakdown in what was becoming a fragile and increasingly threatened social order. Private property rights were under attack. Land was being wrested away from the hands of families who had owned the land for generations. There is no question that for sectors of the Chilean right, the rise and increasing political power of both dedicated social reformists and committed radicals in the mid-to-late 1960s proved traumatic, and this sense of trauma permeates contemporary right-wing narratives of the past.

Through 1990, allusions to politically fraught historical moments abounded. Protesting the recent appearance of brigades of organized political muralists in Santiago, for example, a senator from the traditional right-wing party Renovación Nacional, Francisco Prat, suggested that such brigades would "generate a clear perception to the passersby, and to the citizenry, that under the democratic regime organized groups would once again predominate who, under the umbrella of muralist activity, hide structures, methods, and perhaps equipment that are paramilitary in nature."[15] For Prat, contemporary mural making recalls his uneasy memories of the revolutionary imagery in popular mural making of the 1960s and early 1970s.

Following the president's May 21, 1990, address to the Congress, *Concertación* lawmakers applauded the president for calling the Pinochet regime a "dictatorship"—a term avoided by the major media and former regime supporters—and they called upon opposition senators to recognize their complicity in the "errors" of the past. Christian Democratic senator Carmen Frei, daughter of President Frei (1964–1970), challenged fellow senators from the Right to undergo a "public self-criticism" for the errors of the military regime, just as she claimed the center and the Left had done regarding the Christian Democratic and UP governments.[16] During the same session, socialist senator Nuñez returned to the floor, both to accept that the UP had committed errors, and to deny emphatically that those errors justified what followed:

> None of the errors of the UP—including all those that have been described [by the right]—justifies that in our country people were executed, disappeared, massively tortured; shantytowns, factories and universities were invaded; imprisonment, systematic assassinations . . . No error of the UP justifies that people had their throats slit and that young people were burned alive in Chile. No error of the UP justifies, my fellow Senators, that priests were persecuted in attempts to defame the role of the Catholic Church. It is clear that we committed errors during the Popular Unity government. But none of them justifies the incapacity of the political class—many of them actors standing right here in this room—to have prevented the armed forces from assuming power . . .[17]

The most contentious 1990 Senate debate over the past took place in the immediate aftermath of the discovery in Pisagua, Chile, of bodies of those who had disappeared during the early months of the dictatorship. The discovery shook the country. Clearly the *idea* of disappearances was not unknown among Chileans; virtually from the moment the democratic government was sworn in, Chileans came forward to reveal where bodies were buried. Yet the bodies of these disappeared human beings had been eerily preserved in the northern desert soil and then widely displayed through news footage and photographs. It was new, horrifying, and indisputable.

In the Senate, emotions ran high and *Concertación* senators took rightist senators to task for continuing to defend the dictatorship in light of the evidence from Pisagua. Rightist senators continued to argue that the military had saved the country. To support their claims, Pinochet-appointed senators and senators from the Renovación Nacional and the UDI commonly drew from Christian Democratic leaders' 1973 speeches and letters condemning the UP government and supporting the coup.

After the particularly contentious session on Pisagua, RN senator Sebastián Piñera submitted a proposal for "a new special session, not in order to continue speaking of the past, but rather with the objective of continuing, together, to explore and analyze the best ways to compatibilize in our country the values of truth, justice, peace, and reconciliation."[18] Piñera's use of the term "reconciliation" anticipated the ways in which the Chilean Right would come to appropriate the concept over the course of the 1990s. Nevertheless, in perhaps one of the more confessional-minded narratives of the early 1990s, Christian Democratic senator José Ruiz De Giorgio delivered a lengthy speech entitled, "Reconciliation: The Road toward Democracy," in which the senator distinguished between "my truth" and "other truths." De Giorgio's truth included publicly recognizing that when he first heard the news of the 1986 failed assassination attempt—"real or fictitious"—against Pinochet, the senator felt "disappointed because the dictator had survived."[19] De Giorgio's narrative is sophisticated in its distinction among "truths," memories, and historical facts.

One of the most outspoken rightist Chilean senators was Jaime Guzmán, the leading civilian architect of the Pinochet regime and its 1980 Constitution, as well as the intellectual founder and leader of the UDI. Guzmán consistently trumpeted the military regime's accomplishments, criticized those who termed the regime a "dictatorship," and laced his speeches with thinly veiled warnings against those who might offend the armed forces. Guzmán represented himself as Pinochet's spokesperson, expressing the general's displeasure with the *Concertación* government's claims against his regime. In the days preceding his April 1, 1991, assassination, Guzmán took the lead in rejecting a proposal forwarded by the Renovación Nacional to give Aylwin the power to pardon political prisoners still languishing in jail despite the end of the dictatorship.

It was Pinochet himself who emerged from the hospital to announce Guzmán's death. An angry crowd of Guzmán supporters gathered in front of the Military Hospital to denounce the *Concertación* government as soft on terrorism; some shouted for a return of the DINA, the Pinochet regime's notorious intelligence agency. In the months following the assassination, official discourse focused on the fight against terrorism, understood in the here and now, conducted by non-state agents. Guzmán's assassination traumatized the *Concertación* leadership, and the rightist opposition claimed the discursive high ground.

Guzmán's assassination haunted the efforts of *Concertación* congressional representatives to proceed with several initiatives, from constitutional reforms to a range of actions pertaining to victims of human rights violations. This included short-circuiting any government-led discussion of the Truth and Reconciliation Commission report, released only shortly before Guzmán's murder.

Between 1991 and 1998, elite silence regarding the past endured. There were occasional moments for expressive politics,[20] including such symbolic efforts as those to establish a monument in the general cemetery to the dead and disappeared, monuments to Salvador Allende, and an end to September 11 as a national holiday, yet the efforts were conducted primarily behind closed doors. The call to end September 11 as a holiday would not become significant until 1998, when Pinochet himself endorsed the move. It would take a series of external factors, from the Chilean military's January 2001 public admission of atrocious human rights violations, to the Chilean Supreme Court decision to remove Pinochet's immunity from prosecution, for the leftist governing elite to begin to break the pact of silence.

The ghost of Jaime Guzmán was ever-present in congressional debate over a proposal to erect official statues commemorating the late president Salvador Allende. In contrast to the congressional approval processes for monuments to the late presidents

Jorge Alessandri (1958–1964) and Eduardo Frei, Sr. (1964–1970), which took a mere three months from start to finish, the approval for the Allende monuments took four years. Obstructionist scenes on the floor of the House, particularly, were unusually vehement. Perhaps predictably, efforts to approve the Allende monuments led to tension over how to characterize the UP period, and, in contrast to 1990 congressional debates, the overwhelmingly dominant voices in both the House and the Senate were the Chilean right.

During the House struggle over the monument legislation, RN congressman Juan Enrique Taladriz held Allende and the UP government responsible for the calamities of the 1970s, particularly those whose properties had been expropriated. "Thousands of Chilean families suffered irreparable losses as a direct or indirect consequence of the [UP] government," Taladriz continued. He opposed the statue "in order not to relive that period that is better left forgotten."[21] Echoing rightist historical positions regarding the Truth and Reconciliation Commission, Congressman Carlos Ignacio Kuschel excoriated the Allende government for starting the violence that had wracked Chile, and credited the military regime with averting civil war. True democrats and patriots, he argued, were forced to suspend democracy in order to save it.[22]

A handful of leftist congressmen responded forcefully. "Allende never left this world with blood on his hands!" said socialist congressman Hector Olivares, a former political prisoner and exile who identified himself as a "proud friend" of the late president. "Allende never disappeared citizens from this country nor ordered torture in his government!"[23] In the memory of "el compañero Allende" Congressman Mario Palestro charged that a certain "*olvido*"—a collective forgetting—was taking place in the Chilean transition, a forgetting that would in part be countered by the monuments to Allende.[24]

The one public Senate discussion of the Allende monuments legislation in June 1994 was dominated by several right-wing senators who issued lengthy denunciations of Allende. Twenty-two of the 38 senators present chose to exercise their rights to speak to the law before the vote, and the voices were overwhelmingly Allende opponents, from beginning to end. RN senator Francisco Prat argued that Allende had attempted to lead the country down a violent path toward totalitarianism. UDI senator Hernán Larraín declared Allende's administration "dark, very negative, and, perhaps, one of the worst moments in our history."[25]

While the Right was explicit in its characterizations, congressional members of the Christian Democrats and the center-left Party for Democracy (PPD) were careful not to identify themselves with Allende's leadership. In fact, of the 22 senators who spoke, only four expressed unqualified support for the law. Instead, they argued that Chile would appear "strange" in the eyes of the international community if the senators failed to approve the monuments, given the abundance of streets, plazas, and monuments to Allende throughout the world. PPD senator (and later education minister) Sergio Bitar, a former member of Allende's cabinet and a former prisoner and exile, focused on "tolerance," "diversity," and "reconciliation." Bitar insisted that the role of the Senate was not to be the judge of history.[26]

While the Senate process was less tension-ridden than that of the House, Senate proponents averted a nasty public struggle and potential defeat through an unsavory compromise: in exchange for UDI senatorial support for the Allende monuments, socialist senators had agreed to support an earlier law for monuments to Jaime Guzmán. The socialist senators voted unanimously for the Guzmán monuments. In turn, the UDI senators ultimately voted unanimously for the Allende monuments, assuring the simple majority needed to pass the legislation. The final Senate vote on the Allende monuments was 27 in favor, 8 opposed (all Pinochet-appointed), and three abstentions (all from the Renovación Nacional).

Until 1998, though remnants of historical conflicts continued to echo in congressional debates over contemporary legislation, political elite demands for a return to direct engagement with the historical record largely ceased. This did not mean the past simply went away: on the contrary, human rights groups continued to demand accountability, lawyers representing families of human rights victims continued to press the courts, bodies continued to be exhumed. Every March, on the anniversary of the release of the Truth and Reconciliation Commission Report, and every September 11, on the anniversary of the coup, sectors of civil society organized commemorations and protests. Human rights activists also attempted to focus media attention on violations cases that fell outside the confines of the military-instituted 1978 amnesty law, which covered the large majority of heinous violations and which, until Pinochet's 1998 arrest, Chilean judges consistently upheld. In 1995, the Chilean Supreme Court condemned former military intelligence chief Manuel Contreras and coconspirator Pedro Espinoza to jail for the assassination of Orlando Letelier, former ambassador to the United States under Allende who was murdered by a car bomb in Washington, D.C., in 1976. Nevertheless, political movement with respect to the dictatorship was halting, relegated to civil society and the courts by the executive and legislature.

There were a handful of congressional leaders, such as Christian Democrat Andrés Aylwin and PPD senator Laura Soto, whose lone voices on the past managed to find a place in an editorial or article. Otherwise, there was a long and wide disjuncture between governing leftist politicians, on the one hand, and those sectors of civil society for whom the past would not be forgotten, on the other. The governing Left moved to a defensive mode.

Breaking the Silence

The year 1998 marked a turning point, in which a range of memory-laden political moments would return history to the fore. For the political elite, expressive politics began with Pinochet's retirement from the military and subsequent entry into the Senate. A handful of congressmen attempted to challenge the legality of Pinochet's admission to the Senate, and senators and congressmen staged a silent protest the day Pinochet entered the Senate floor. Expressive politics continued with a series of reflections and debates linked to the twenty-fifth anniversary of the coup, and public debate among the leadership peaked with a series of events triggered by Pinochet's arrest.

While the Chilean foreign ministry warned Pinochet that traveling to London subjected him to the risk of arrest, the former dictator was unconcerned. He had traveled to London in the 1990s, during the conservative Margaret Thatcher administration. That, however, represented a different political moment. Britain's prime minister was now Labour Party leader Tony Blair. Moreover, in 1996, Spanish magistrates Baltazar Garzón and Manuel García Castellón had begun investigations into the forced disappearances, executions, and torture of Spanish and non-Spanish nationals in Chile under the Pinochet dictatorship. Garzón issued an international warrant for Pinochet's arrest.

On the evening of October 16, 1998, while Pinochet was recuperating from spinal hernia surgery in a London clinic, Scotland Yard officials entered Pinochet's room and pronounced him under arrest. For the next year and a half, human rights lawyers, lawyers for Pinochet, and Chilean government officials, battled over Garzón's request for Pinochet's extradition to face trial in Madrid for genocide, terrorism, and torture. Ultimately, British home secretary Jack Straw would side with the Chilean government's demand that Pinochet be returned to Chile, potentially to face Chilean courts. Straw publicly based his decision upon humanitarian, medical grounds, which, importantly, did not claim the former dictator was free from culpability.

Pinochet's 18-month-long absence while under house arrest outside London opened new space in Chile for exploration of the past. Increasing numbers of Chilean judges began to reinterpret the 1978 amnesty law, declaring that "disappearing" a person constituted a "permanent crime" and therefore was not subject to the amnesty law. For the first time, the four branches of the Chilean military (the army, navy, air force, and police) agreed to a dialogue with lawyers for the families of the executed and disappeared.

In another vein, a group of sons and daughters whose parents were murdered and disappeared under the dictatorship organized Acción, Verdad, y Justicia (Action, Truth, and Justice). The group took to the streets with a banner, posters, and information that celebrated Pinochet's arrest and reminded society that their parents had suffered in concentration camps and torture centers, and that many persecutors continued to enjoy impunity. "Our forays were celebratory, but they were also to denounce" such crimes, stated a member of the group. Acción, Verdad, y Justicia's street acts were both colorful and captivating, and the group sensed a society receptive to its message and concerns. Representatives of the group met with a member of the Argentine human rights organization HIJOS, similarly composed of children of the disappeared, whose main objective is to expose former persecutors and to continue to pursue investigations of the whereabouts of their loved ones. The HIJOS representative convinced the Chile group that similar actions could be taken in Chile.

The Chile group has become known as the La Comisión Funa, or La Funa, a term that some claim derives from a Latin American folkloric practice of publicly "outing" a community debtor. Today Chilean young people use the term, also popularized by Chilean musician and songwriter Joe Vasconcellos, to describe someone who has lost all credibility.

Since its founding, La Funa has conducted dozens of "*funas*," where members of the group gather, often before a former persecutor's home or office, to out the repressor. La Funa carefully selects these outings, based on a range of information garnered from human rights archives, court documents, and testimonies. The group's first *funas* involved two or three dozen participants who unfurled a large banner and collectively read the charges against the accused victimizer. Since the early *funas*, the group has attracted as many as several hundred participants at an action.

La Funa clearly represents a direct linking of young people with collective memories of dictatorship and repression. The organization draws its identity from the human rights movement and from life histories of leftist militants and La Funa members' own families who were murdered by the military regime. The *funas* have outed both those persecutors involved in the early years of massive repression against leftist activists and those who participated in the more select tracking down, torture, and murder of leftist militants who took up arms against the dictatorship in the 1980s. La Funa's work challenges Chileans to remember a broader, more nuanced set of accounts of repression and resistance. The group's mission is also to contribute to Chilean society's overcoming its fears of denouncing the powerful for wrongdoing.

Congresswoman Isabel Allende, daughter of the late president, was unafraid to express her frustrations regarding her colleagues' fear of conflict:

> . . . There are still very real fears. And anything which appears as conflict is something seen as very worrisome. In principle, all of us want to avoid conflict, but there are times when trying to avoid conflict will only create more conflict. Today we are in absolute denial of conflict, which creates a kind of consensus that in part is superficial. There is a very real part of the population that remains traumatized. The whole theme of conflict, violence, is very worrisome for them . . . So our challenge at this moment is how as the government to move forward toward the next presidency united but getting rid of this tendency to create scary ghosts . . . We have to break this pattern.[27]

On March 18, 2003, Isabel Allende became president of the House of Representatives. Though her victory was narrow, the election of a representative who not only bears Allende's name but also insists on the importance of confronting the past symbolizes a renewed willingness among Chilean politicians to explore their traumatic history. In addition to Isabel Allende, there are members of the governing left who have taken a more proactive public stance regarding the need to come to terms with the past, particularly in challenging dominant narratives and imagery of victors and vanquished.

Socialist senator Carlos Ominami—the son of a military general who was arrested, tortured, and exiled by the dictatorship—has clearly been uncomfortable with the governing left's silences; his discomfort visibly surfaced with Pinochet's March 1998 self-appointment to the Senate. Ominami himself is a former revolutionary MIR activist who adopted the son of MIR leader Miguel Henríquez, who died in a shoot-out with the Chilean military in 1974. Many of the disappeared are Ominami's former comrades.

In the immediate aftermath of the military's admission that officers threw over 100 bodies of the disappeared into the sea, Ominami rose to the Senate floor to deliver a powerful address challenging rightist members of the Senate to admit to their historic complicity, to accept the past military regime's "machinery of death":

> I am aware that some ex-functionaries of the regime intervened to save people who had been arrested. I am aware, even, that someone as important to the Senators and Congressmen of the Opposition, like Jaime Guzmán, intervened to rescue some detained from the hands of the DINA [Chilean military intelligence]. But these

interventions are also a demonstration that what was happening was known and that it was necessary to rescue those friends from this machinery of death and from making people disappear.[28]

Ominami insists upon the imperative of asserting alternative "historic reference points." Yet like Adriana Muñoz in the House, Ominami doubts the Left's willingness and capacity to engage in history and memory battles given the political correlation of forces. While Ominami never publicly disputed the *Concertación's* major efforts to return Pinochet from London rather than face extradition to Spain for trial, the 2002 Chilean Supreme Court decision to declare Pinochet mentally unfit to stand trial left Ominami with a curious comparison:

> If you compare us with Spain and the Spanish left, remember that Franco died in his bed, and this is something not too glorious for the left, and in a way, that is what is happening here.
> . . . You can see it was an industry of disappearing, 500, 1,000 people, backed by a consensus, as well as an infrastructure and a language . . . How could it be that when the armed forces said, "We threw people into the sea," that there wasn't a dramatic, explosive outcry, from everyone? You realize they just found the remains of thirteen, who had been in La Moneda [the Presidential Palace, serving as Allende's bodyguards], whom the armed forces claimed had been thrown into the sea, which wasn't true, either.[29]

The remains of thirteen represent several of Ominami's former comrades who had served as Allende's personal bodyguards. While Ominami's account represents a rationalist/political learning explanation of why themes of the conflictive past do not surface, the tone of the interview revealed anger and disgust, founded upon memories of lost comrades and a responsibility to their memories given his own survival.

Senator Ricardo Nuñez has spent the years since the democratic transition of 1990 facing fellow Senate members who include former Pinochet civilian ministers and high military officials. On the Senate floor, he has challenged both general and specific rightist accounts of the UP period, the military coup, and the repression that followed. Nuñez has observed first-hand how the Chilean right continues to deny culpability for the errors and atrocities of the past, and has concluded that the Right will not amend its interpretation of the past:

> Chile is a country that denies its history. It is a kind of syndrome of neither accepting nor reconstructing its past, because for the right, as for the business class, and the military, and for the left in its failed state, all of us see the past as an instance in which we all committed errors.
> . . . We will always have a right that never accepts its responsibilities for the past, that will always explain the past in terms of a natural process, that denies the real strength of the left, that there was an Allende, a Communist Party, and all this goes beyond merely a question of the Cold War, or the US, a US that clearly has an enormous responsibility for the past as well. They consider abnormal or they deny that we had historically the largest Communist Party, together with the

Italian Communists, outside the Soviet Bloc, of any country in the world, and that our Socialist Party was socialist, not social democrat, a revolutionary socialist force. We were an anomaly, and we had no guerrilla force. And we possessed a massive popular project. The right considers this an aberration that just could not have occurred . . .

We will always have this intimately cultural syndrome, and that is the double reading of our history. This is something we've always had . . . and from now on for many, many years to come we will have this double reading of the dictatorship.[30]

On August 16, 2000, a full decade after Nuñez's request that the Senate engage in a special session to explore the 1970–1973 years, the rightist opposition proposed and was granted a special session on "The Causes of the Existing Situation in Chile on September 11, 1973." The political context of the request, however, was dramatically distinct from that of 1990. The Right deemed the session a necessary response to Pinochet's loss of immunity from prosecution, a Supreme Court decision that floored Pinochet sympathizers.

The Senate session proved to be the lengthiest senatorial hearing on the pre-1973 period since the transition from authoritarian rule. While the Right exercised more time on the floor than senators from the *Concertación*, the debate included detailed accounts from across the political spectrum, addressing an array of organizations, events, and positions from the pre-1973 years. The Senate session affirmed Nuñez's claims regarding rightist interpretations of history and a double reading of the past. Yet the session also symbolized the Left's breaking the silence regarding the past.

What has occurred to change the ways the governing left publicly identifies with its past? It clearly took Pinochet's London arrest, but perhaps more importantly, it took the Chilean military's confessions regarding throwing bodies into the sea, and the Chilean judiciary's removal of congressional immunity from civil prosecution for Pinochet. In other words, it took the very institutions that had been complicit with past atrocities, that had stonewalled societal demands to address the past, to bend, to crack, even if just a little, and even if, in the case of the military, it was in an attempt to save itself from mounting human rights violation charges. A new generation of civil rights-minded judges has facilitated the process. Since 2000, over 300 former and active duty military officers have been subpoenaed for human rights violations, and dozens have had to serve time, albeit limited, behind bars.

In 2003, the *Concertación* administration proposed major legislation calling for an expanded investigation into the whereabouts of the disappeared; for the first time, the government established a commission to define and recommend compensatory measures for the many thousands of victims of torture who had survived. In addition, Chile's Council on National Monuments approved a civil society-initiated proposal to establish the National Stadium, which served as a notorious holding facility for political prisoners during the early months of the regime, as an historic monument to commemorate those who were imprisoned, tortured, and killed there.

Individual and collective memories of a turbulent, traumatic past play a crucial role in post-authoritarian politics. Recent developments—often in spite of rather than because of Chilean political leaders—have laid bare the ways in which traumatic historical legacies can dramatically jolt as well as doggedly plague democratic regimes.

It would seem that Chilean politicians across the political spectrum have now firmly and publicly accepted the inevitability of the unearthing of the traumatic past. Political elite debate seems finally to recognize a need, therefore, to come to terms by providing more comprehensive and systematic political mechanisms to address, and even embrace, the search.

Notes

1. Interviews with Sarah Shanley, Santiago, July 4, 2002.
2. "87% de los chilenos opina que aún no se logra la reconciliación," *El Mercurio Online,* July 4, 2003.
3. See Manuel Antonio Garretón, "The Feasibility of Democracy in Chile," *Canadian Journal of Latin American and Caribbean Studies,* 15: 30 (1990); and Peter M. Siavelis, *The President and Congress in Postauthoritarian Chile: Institutional Constraints to Democratic Consolidation* (University Park, PA: Pennsylvania State University Press, 2000).
4. Interview with Carlos Ominami, July 1, 2002.
5. For a detailed and nuanced analysis of Chilean political learning, see Manuel Antonio Garretón and Malva Espinosa, "Chile: Political Learning and the Reconstruction of Democracy," in Jennifer McCoy, ed., *Political Learning and Redemocratization in Latin America: Do Politicians Learn from Political Crises?* (Miami: North-South Center Press, 2000), 37–71. See also Nancy Bermeo, "Democracy and the Lessons of Dictatorship," *Comparative Politics* 24 (April 1992), 273–291.
6. Interview with Ricardo Nuñez, June 25, 2002.
7. Interview with Palma, November 8, 1991.
8. Alexander Wilde, "Irruptions of Memory: Expressive Politics in Chile's Transition to Democracy," *Journal of Latin American Studies,* 31: 2 (May 1999), 476.
9. Interview with Adriana Muñoz, July 1, 2002.
10. Interview with Patricio Rivas, November 21, 1991.
11. Interview with Patricio Rivas, June 12, 1998.
12. Mariano Ruiz-Esquide, Senate, Session 9, April 25, 1990, 340–341.
13. Ricardo Nuñez, Senate, Session 9, April 25, 1990, 353–354.
14. Eugenio Cantuarias, Senate, Session 11, May 8, 1990, 437–438.
15. Francisco Prat, Senate, Session 4, April 3, 1990, 193.
16. Carmen Frei, Senate, Session 1, May 29, 1990, 56–57.
17. Ricardo Nuñez, Senate, Session 1, May 29, 1990, 83.
18. Sebastián Piñera, Senate, Session 10, June 20, 1990.
19. José Ruiz De Giorgio, Senate, Session 13, July 3, 1990, 964.
20. This concept is taken from Wilde, "Irruptions."
21. Juan Enrique Taladriz, House of Representatives, Session 2, June 2, 1992, 43.
22. Carlos Ignacio Kuschel, House of Representatives, Session 2, June 2, 1992, 46.
23. Hector Olivares, House of Representatives, Session 2, June 2, 1992, 47.
24. Mario Palestro, House of Representatives, Session 2, June 2, 1992, 41–42.
25. Hernán, Larraín, Senate, Session 5, June 14, 1994, 512.
26. Sergio Bitar, Senate, Session 5, June 14, 1994, 502.
27. Interview with Isabel Allende, June 12, 1998.
28. Carlos Ominami, Senate, Session 21, January 9, 2001, 2322.
29. Interview with Carlos Ominami, July 1, 2002.
30. Interview with Ricardo Nuñez, June 25, 2002.

Suggestions for Further Reading

Constable, Pamela and Valenzuela, Arturo, *A Nation of Enemies: Chile Under Pinochet* (New York: W.W. Norton, 1991).

Corradi, Juan, Fagen, Patricia Weiss, and Garretón, Manuel Antonio, eds., *Fear at the Edge: State Terror and Resistance in Latin America* (Berkeley: University of California Press, 1992).

Davis, Madeleine, ed., *The Pinochet Case: Origins, Progress and Implications* (London: Institute of Latin American Studies, 2003).

Hite, Katherine, *When the Romance Ended: Leaders of the Chilean Left, 1968–1998* (New York: Columbia University Press, 2000).

Hite, Katherine and Cesarini, Paola, eds., *Authoritarian Legacies and Democracy in Latin America and Southern Europe* (South Bend, IN: University of Notre Dame Press, 2004).

Jelin, Elizabeth, *State Repression and the Labors of Memory* (Minneapolis: University of Minnesota Press, 2003).

Kornbluh, Peter, *The Pinochet File: A Declassified Dossier on Atrocity and Accountability* (New York: The New Press, 2003).

Wilde, Alexander, "Irruptions of Memory: Expressive Politics in Chile's Transition to Democracy," *Journal of Latin American Studies*, 31: 2 (May 1999), 473–500.

CHAPTER 4

Political Uses of the Recent Past in the Spanish Post-Authoritarian Democracy

Carsten Jacob Humlebæk

The Spanish Civil War (1936–1939) pitted supporters of Spain's democratic Second Republic, the "republicans," who received aid from the Soviet Union, against right-wing rebels led by General Francisco Franco and backed by Nazi Germany and fascist Italy, who called themselves "nationals." The war produced a profound fracturing of the national community that was intensified by the dictatorship installed by Franco after defeating the republicans. The leitmotiv of the dictatorship was "Spain: One, Great and Free" (*España: Una, Grande y Libre*). In a country with strongly pronounced regional identities and separatist movements, "One" referred to a centralist administrative structure and the absence of all demands by regional separatists known as peripheral nationalists, such as the Basque, the Catalan, and other groups that speak a language other than Castilian Spanish and seek autonomy or independence from Madrid. "Great" invoked the heroic connotations of the Spanish imperial age. Finally, "Free" did not mean liberty, but rather national sovereignty with reference to the resistance against attempts of foreign powers to intrude into Spanish politics. By establishing an official nationalism that excluded the losers of the civil war, Franco's regime deprived approximately half the citizens of their identity as Spaniards. This explains the urgent necessity of reconstructing the nation on a basis of reconciliation after the death of Franco in November 1975. The post-Franco change of regime took the form of a smooth, gradual, and relatively rapid reform process, which was characterized by concord among most of the parties involved. The formula employed was to dismantle the Francoist regime from within, using its legal and parliamentary structure against itself. The culmination of this process was the consensual writing of a new democratic constitution and its approval in a referendum by ample majority in December 1978, just over three years after the death of the dictator.

The character of the Spanish transition from authoritarian rule, and to a large extent also its success, was intimately related to specific ways of dealing with the traumatic past, which was part and parcel of an attempt to construct a coherent account of the nation. When Franco died, Spain was not only "haunted" by the memory of the dictatorship, but also by the still vivid memory of the 1930s. The traumatic memory of the Second Republic (1931–1936) and the civil war had been carefully shaped and instrumentalized by the dictatorship. Victory in the civil war constituted the founding myth and principal source of legitimacy of the Franco regime. But from the early 1960s onward, the discourse on victory was phased out in favor of a discourse on peace, stability, and economic development that stressed the achievements of the Franco regime. This anti-revolutionist discourse emphasized, above all, the importance of not repeating the civil war, the "never again civil war" discourse. Significantly, the losing side in the war, the oppressed opposition, also to a large extent favored the same interpretation of the historical experience. The Francoists in power were anxious to legitimate the regime, and the opposition was anxious to avoid a repetition of their experience of war, exile, and repression, so both agreed on the importance of not repeating the civil war.

After the death of Franco, it was difficult to use any discourse of Spanish nationalist affirmation due to its monopolization by his regime. For any Spaniard who had lived under the dictatorship, expressions such as "Spain," the "Spanish Nation," or cheering "Viva España" immediately evoked Francoist discourse. During the civil war the Francoists named themselves "the nationals" in their claim to be the true and legitimate representatives of the Spanish nation. Partly as a result of this dynamic, the left wing developed a problematic relationship with the Spanish nation. The Francoist dictatorship repressed not only the supporters of the republic, but also the manifestations of the subnational cultures of Spain. This explains the paradox that most Catalan and Basque nationalist political forces, even if conservative, were close to the left-wing opposition parties in exile. The attainment of democracy and regional autonomy by the so-called historical nationalities became inextricably linked. On the one hand, peripheral nationalists supported democratization together with the rest of the democratic opposition and, on the other, the leftist parties supported devolution and decentralization to the point that they were dissuaded from adopting any "national" discourse regarding Spain.

In the mid-1970s, both the Spanish Communist Party (*Partido Comunista de España*, or PCE) and the Spanish Socialist Workers Party (*Partido Socialista Obrero Español*, or PSOE) defended "the right to self-determination" of the historical nationalities of Spain. As a consequence of the community of interests between the left wing and the peripheral nationalists, the very name of the country was often avoided by the Left during the post-Franco transition. Instead paraphrases such as "this country" or "the Spanish State" were, and still are, the preferred ways of referring to Spain if one wants to avoid being associated with the excesses of the dictatorship. This reflects apparent doubts within the left wing about the exact definition of Spain and the nation, and what the relationship between these two entities might be. For the right wing there was no doubt, the nation was Spain, period. Therefore, although almost all parties agreed on the value of democracy and on the relation between democracy and some kind of regional autonomy, they did not agree on a definition of the nation. The central problem thus became how to connect democracy and national discourse.

After the death of the generalísimo, the "never again" discourse, which was the result of a particular historical memory, guided the majority of political actors toward moderation and consensus in order to refound the nation, and to conceptualize it in an inclusive and democratic way. The fear of a return of the civil war scenario thus became the basis for the construction of social peace and reconciliation, which was translated into tangible politics, for example, by a series of decrees and laws that gave amnesty to political prisoners from the opposition against the Franco regime. But the initiatives of reconciliation that were carried out by the new regime not only concerned the opposition against the dictatorship; they also broadened the amnesty to include the infringements committed during the Franco regime. This was necessary for the operationality of the transition. If the idea of consensus worked as a guiding principle when it came to constructing a new democratic conception of the nation, the other half of the basic social pact behind the transition concerned the way of dealing with the authoritarian past. It became imperative to focus on the future and silence any questions about guilt and responsibility. Much later, this common agreement was denounced as a "pact of silence," a point to which I shall return below, but in its contemporary context it provided benefits for both parties. In return for letting the regime move toward democracy and parliamentary monarchy, thus giving up their power, the Francoists achieved legal amnesty against the possible penal consequences of their actions during the dictatorship and thereby avoided the revenge of those that they had oppressed. The second in command of the PSOE at the time of the transition, Alfonso Guerra, recently admitted that "in the transition, we forgot about what the right wing had done on the condition that it did not repeat the same behavior."[1] In return they gained a democratic system, their longed-for freedom, amnesty for their political prisoners, the granting of pensions to the republican soldiers and their widows, restitution of part of the confiscated property, and so forth, all of which were initiatives that concerned injustices committed by the Francoist side. In reality, however, the execution of these measures of reconciliation was rendered very difficult or impossible by bureaucratic practices and interminable delays. Before issuing pensions, for example, the authorities demanded documents that in many cases never existed or were destroyed during the war. Likewise, the restitution of confiscated property was limited essentially to a part of the property of political parties and unions, and was hampered by the same demand for nonexistent documents. Individuals were never compensated for their confiscated property.

The fact that the Francoist instrumentalization of the memory of the civil war focused on that period as a trauma, and that it was experienced as such by both victors and vanquished, determined that the sole fact of establishing a consensual democratic regime without violence was experienced as a process of mastering the past. Being able to agree on the basic "rules of the game" was interpreted as the overcoming of a historical problem by both sides. But it was not the *lived* past that was being mastered; the process was restricted to mastering only the past in mythicized form, as portrayed by the Francoist legitimization discourse, that is to say, through the myth of the ungovernable character of the Spaniards. The historical problem of the trauma of the civil war, then, was overcome only in that a consensual democratic regime was established without violence. But the lived past in the form of individual loss was not addressed. When it came to openly confronting the past and discussing questions of historical justice, guilt, and responsibilities, the transition years (1975–1982) only to a very limited extent resembled the German process of mastering the past [*Vergangenheitsbewältigung*], which was often held up as a model. The main civilian and military institutions of the Franco regime remained intact, criminals were not punished, public employees were not purged, the history of the dictatorship was not a public issue, and most victims were rehabilitated only indirectly or very late. The amnesty of both political prisoners and perpetrators satisfied the rehabilitation of the victims while obstructing the punishment of perpetrators and the purging of officials. There was thus no de-Francoization of the military or other institutions equal to the denazification of Germany after World War II.

What the amnesty law silenced in terms of penal responsibilities was mirrored in the political realm by the informal pact not to instrumentalize the past politically. The Francoist past was largely silenced in an attempt to avoid destabilizing ideas of revenge and collective guilt, which was believed to be necessary to achieve the longed-for reconciliation of the two sides. A democratic future for Spain and a profound political and public debate on the dictatorship were perceived as conflicting goals. Instead of *Vergangenheitsbewältigung* in the German sense, in Spain the past was kept under tight control through a series of political practices such as "never again," moderation, consensus, and focus on the future.

This agreement was later disrespectfully called the "pact of silence." The pact, however, did not have the same degree of force in all domains. The past, particularly the civil war, was actually very present in the cultural and social domains, and the "pact of silence" was principally limited to the political sphere. Moreover, the distinction between the memory of the civil war and that of the Franco regime is important, since these generated different levels of consensus. While the pact of silence with regard to the civil war was based on equilibrium between the parties, since both sides had taken part in the atrocities of the war, the pact was not characterized by the same harmony insofar as it concerned the Franco regime. The benefits of the accord were not equally distributed since those who had collaborated actively with the dictatorship had more to gain than others who had opposed it. By agreeing not to instrumentalize the past politically, the former opposition limited its own possibilities of obtaining political gains from its past.

This fact made the pact fragile in the long run and explains the rupture of the pact in the mid-1990s. Until then, the PSOE generally respected the pact, and thus never tried to pass an official condemnation of the military uprising of July 1936 or the Franco regime in parliament during the four legislative periods it governed (1982–1996). This reticence was dictated, at least during the early years of the PSOE government, by the wish to consolidate the democracy. Once democracy had been consolidated, however, the socialists could have risked breaking the pact for political gain, but they did not. On the other hand, there was no pressure on them from other sides to do so. The only public statement by the PSOE government on the civil war was the official declaration in 1986 on the fiftieth anniversary of the uprising. "Due to its fratricidal character," the declaration read, "a civil war is not a commemorable event."[2]

The memory of the civil war had been kept alive by the Franco regime in order to legitimize itself by arguing that Spaniards had an underdeveloped political culture.

By 1986, the PSOE government held that the political culture of the Spaniards was more or less equal to that of other developed democratic countries, and that, as the declaration continued, "the Civil War . . . is not—nor should it be—a living present." Spaniards, in other words, were morally obliged to remove the civil war from the present and let it become history. The government then paid explicit homage to both sides of the war, recognizing that each fought for its idea of a better Spain. At the same time, however, the government also identified itself with the inheritors of "those who fought for liberty and democracy in Spain," that is to say, the republican side. Current Spanish society, according to the government, had shown its will to forget the wounds of the civil war and to use the memory of it only to avoid permitting "the ghost of war and hate" to dominate Spain again. While explicitly mentioning the recent past might seem to be breaking the pact, it was clear that the government did this only to confirm the irrelevance of this past to the present, thus ratifying the pact not to instrumentalize the past politically.

The pact was truly broken by the PSOE first in the election campaign of 1993, in which they, for the first time, feared losing to the conservative Popular Party (*Partido Popular*, or PP). The Socialist Party elites therefore decided to focus their campaign around an instrumentalization of the Francoist past of the PP, insinuating that democracy would be endangered if the latter won the elections. This strategy proved efficient and the PSOE won the elections, but returned afterward to its previous practice of respecting the pact, for example, by not attempting to officially condemn the civil war or the Franco regime.

Within the political sphere, then, discourse on the civil war and the dictatorial past thus remained almost absent during the first two decades of democracy. By the early 1990s, the traumatic past of the civil war and the Franco regime seemed to have become pages in the history books, as the declaration of 1986 called for, thus ceasing to cause problems for the politics of the present. Democracy had been consolidated, and Spain had become a full member of the European Economic Community, thereby ceasing to be an exception in Western Europe. This new Spanish identity was manifested internationally in 1992 through the multiple celebrations in Spain of the Olympic Games in Barcelona, the Universal Expo in Seville, Madrid as cultural capital of Europe, and the quincentennial of Columbus' discovery of America. Not without reason, 1992 was baptized "the year of Spain."

Shortly thereafter, however, and particularly after 1996, when the conservative PP came to power, political debates on the civil war and the Franco regime began appearing with a certain frequency. It was as if a wider public suddenly became aware of the existence of some kind of pact to silence or forget the past, and it became commonplace among analysts of the transition to democracy to denounce "pacts of forgetting" or "collective amnesia." This development was in fact the first sign that the pact not to instrumentalize the past politically was breaking up. The political uses of the recent past were changing. One of the first of these occasions was the celebration of the sixtieth anniversary of the establishment of the International Brigades in November 1996.

The Brigades were composed of volunteers (*brigadistas*) from many countries who fought against the fascist forces in the civil war. Juan Negrín, the socialist prime minister of the Second Republic's last government until overthrown by Franco, had promised the *brigadistas* Spanish citizenship when the International Brigades were dissolved and sent home in October 1938. For the first 20 years after the death of Franco, however, nothing happened despite several requests. But in November 1995, a proposal presented in Parliament by the Basque Nationalist Party (*Partido Nacionalista Vasco*, or PNV), the left-wing party United Left (*Izquierda Unida*, or IU), and the PSOE to fulfill Negrín's promise and confer Spanish citizenship on the members of the International Brigades was voted unanimously. Consequently, in November 1996, on the sixtieth anniversary of the establishment of the Brigades, approximately 370 aging former *brigadistas* arrived in Spain to receive their Spanish citizenship. This positive commemoration of a particular element of the losing side in the civil war was a clear change from the earlier line maintained by the PSOE. If, in 1986, the governmental declaration stated that the civil war should not be a living present, the symbolic conferring of Spanish citizenship on the *brigadistas* a decade later transformed the value of their sacrifice in the 1930s into a living present of the mid-1990s. It actually meant extending the nation to include those volunteers, whom the Francoists had always regarded as vanguards of a communist revolution threatening Spain.

Between the vote in Parliament and the actual visit, however, the government had changed to José María Aznar's PP, which in spite of being—at least formally—in favor of conceding the citizenship, was much less enthusiastic about the planned ceremony in Parliament and the various celebrations across the country. During the debate in Parliament the year before, all the other parties had referred to the conferring of citizenship on the *brigadistas* as a historical debt and a pending cause of the young Spanish democracy. The representative of the PSOE had explained the delay by referring to the informal pact not to instrumentalize the past politically, saying that to consolidate democracy a certain measure of silencing had been necessary. The representative of the PP, however, used his entire speech to explain the juridical technicalities of the citizenship issue and the democratic credo of his party, avoiding any mention of the International Brigades directly. On behalf of his group, however, he also admitted the need to heal the wounds of Spanish society and show an example of cohabitation [*convivencia*] and responsibility for future generations.

At the actual ceremonies in 1996, however, practically nobody from the high ranks of the PP was prepared to actually meet the *brigadistas*. Both Prime Minister Aznar and the then president of Parliament, Federico Trillo, excused themselves from participating in the central citizenship ceremony. King Juan Carlos did not appear either, in spite of his frequent appeals for reconciliation. According to the protocol, responsibility for leading the ceremony then fell to the first vice president of Parliament, Enrique Fernández-Miranda from the PP. He tried to reduce the magnitude of the event by preparing a reception for a delegation of only ten representatives of the *brigadistas*. The veterans, however, succeeded in insisting that all be received in Parliament, and as a consequence Fernández-Miranda also excused himself. In the end, therefore, it was the second vice president of the Parliament, Joan Marcet from the PSOE, who presided over the official ceremony on November 6, 1996. It was thus clear that the PP, although in favor of conferring citizenship, would have preferred that it be done in silence without talking about the civil war or the heroism of the International Brigades. The mere presence of the *brigadistas* supposed an irruption of an unpleasant historical memory in the midst of Aznar's so-called second transition, by which he referred to a recovery of the consensus of the early post-Franco transition. Their arrival meant a rupture of the pact of silence that Aznar was seeking to renew.

There were many other official acts in honor of the *brigadistas* all over Spain, the most important ones in Guernica, Valencia, Barcelona, and Albacete, as well as in the towns around Madrid. At several of these events the disregard of the PP toward this symbolic measure was repeated. The Catalan Parliament, for instance, celebrated a plenary session in honor of the *brigadistas* in which all the political forces participated except the representatives of the Catalan PP. In Albacete, where the city council was governed by the PP, the mayor, Juan Garrido, did not attend the commemoration during which a monument to the International Brigades was inaugurated by the president of the region Castilla—La Mancha, José Bono of the PSOE. The mayor considered that single official reception in their honor, that is to say the one celebrated in the national Parliament, was enough.

The visit of the *brigadistas*, and the generally reluctant attitude of the PP to participate in honoring them, sparked a nationwide debate. The president of Parliament, Trillo, justified his absence from the ceremony saying: "it is better . . . not to resuscitate sad events from Spanish history."[3] This position, according to which the civil war is best left in the pages of history, followed the position of the PSOE government from 1986 that the civil war should not be "a living present." But where the governmental declaration of 1986 made a point of distinguishing between the two sides of the conflict, the position of the PP government ten years later avoided making a distinction between the sides, labeling the war as a "sad event." This position was directly inherited from the interpretation of the conflict prevailing during the late Franco regime that emphasized the guilt of all involved parties. At the beginning of its first term in government, then, the Popular Party showed a clear preference for silence with respect to the political use of the recent past.

It is worth noting that in this question the left wing and the peripheral nationalists of the Basque country and Catalonia usually sided with each other against the right wing, just as they had done during the Franco regime and in the transition years. The fault line in Spanish politics with regard to the politics of the past thus was not simply a division of left and right but a complex one determined by certain shared memories.

One of the first attempts to achieve an official condemnation of the civil war and the Franco regime was inspired by the sixtieth anniversary of the end of the war in 1999, which the opposition parties wanted to officially commemorate as the sixtieth anniversary of the exodus of Spanish Republicans into exile. Inspired by an institutional visit of the Commission of Foreign Affairs to Mexico, where the members of the commission to their surprise discovered that the sixtieth anniversary of the Spanish exile was being publicly and officially celebrated, in September the opposition filed a proposition to do the same in Spain. As in the case of the *brigadistas*, this initiative corresponded to a will to distinguish between the parties involved in the civil war and positively commemorate a particular part of the losing side. The PP filed a counter-proposition in which it resisted distinguishing so clearly between the sides. The proposition of the opposition parties was actually voted by a majority, since the PP during the legislative period 1996–2000 did not have an absolute majority, and the parliamentary opposition has taken several similar initiatives since then.

One of the reasons for these repeated initiatives is to be found in the kind of retroactive justice that was used in the Spanish transition. At no point in time was a "truth commission" along the lines of, for example, the South African model an option. Such an idea was contrary to the core agreements behind the transition, and it was made impossible by the 1977 Law of Amnesty that erased all prior political and penal responsibilities. Instead, a series of limited legal measures toward moral and economic restitution was taken as the political climate cooled down, which amplified the scope of the basic measures taken during the early transition years. This system of bit-by-bit rectification has, however, obliged the parliamentary representation to repeat this debate on silencing and forgetting every time a new victim group to which the existing measures had not done "historical justice" enters the public sphere.

Generally, the initiatives were rhetorically opposed to what was described as a process of forgetting and historical amnesia. The awarding by the government of the Great Royal Cross of Civil Recognition (*Gran Cruz Real de Reconocimiento Civil*) to Melitón Manzanas, a senior official in San Sebastián, as a victim of terrorism, for example, provoked the Basque nationalists to begin another parliamentary debate in February 2001 aimed at condemning the military uprising of July 1936. The terrorist organization Basque Homeland and Freedom (*Euskadi Ta Askatasuna*, or ETA) had killed Manzanas, but the objection to awarding him the Cross was that he had been a torturer and executioner in the Franco regime and had tortured many Basques during his time in office, which the opposition now accused the government of obscuring. An IU deputy declared:

> [T]he forgetting, the oblivion, the attempt to reinvent, the attempt to rewrite, at the margin of this philosophy of reconciliation in the construction of which we all participated during the decade of the 1970s . . . [constitute] a reinvention of history that we do not intend to accept silently; forgetting as policy, amnesia as strategy.[4]

This amnesia was interpreted as a corruption of the spirit of reconciliation that had dominated the transition without recognizing that the pact not to instrumentalize the past politically was a fundamental part of the broad political consensus behind the transition. According to the IU, amnesty had been turned into amnesia by the PP, as if "historical amnesia" was something invented by the PP alone. In another plenary debate of May 2001, on the rehabilitation of the anti-Francoist guerrilla force, the so-called Maquis, the proponent IU thus stated that a moral obligation to break the circle of amnesia was incumbent on the members of Parliament: "we would take an important step forward towards breaking the forgetfulness, the oblivion . . . . The transition is searching for the last entrenchments of forgetfulness."[5] Implicitly or explicitly, what was presented as a corruption of the original "spirit" of the transition was often called an act of calculated rewriting of history to promote a political project, which was incompatible with the transition. On these occasions, it was not acknowledged that there was a continuity between the political use of the past practiced by the PP and the attitude maintained by the PSOE during its prolonged government. The criticism of the PP was also related to its absolute majority in Parliament, as the deputy already quoted continued: "an absolute majority cannot rewrite history, it ought not to be tempted [to do so]."

At other times, an understanding was expressed that the transition process made it impossible to avoid a certain silencing. But after a quarter of a century, it was believed, Spanish society should be mature enough to complete the process. The distance in time with respect to the transition and the changed political culture ought to make it possible to rectify these pockets of unjust forgetting. A change in the political use of the recent past was thus perceived as possible and necessary. This manner of viewing things was manifested, for example, in a debate in June 2001 on the revision of sentences and moral restitution of Spaniards who were executed for political or moral motives. This was promoted by the IU: "the peace of mind, the serenity, the tranquillity that these 24 years give us should permit us to attempt to accomplish the conclusion, which by no means implies forgetting, of this grey and sad page in the book of Spanish history."[6]

It is equally clear, however, that these initiatives also were motivated by an element of political tactics. The opposition parties were attempting to put the conservatives in a difficult position. It was well known that some of the former supporters of Francoism had voted for the PP. The strategy of the opposition was to either force the PP to publicly condemn the dictatorship and thereby disappoint their rightist voters, or to stigmatize the party as "Francoist" if it eventually decided not to condemn the dictatorial past.

The Popular Party invariably refused to support the various initiatives, sometimes creating its own counterproposals. Generally, the denial was characterized by an unwillingness to analyze the civil war and distinguish between the parties involved. Instead, the PP preferred to follow the credo of the late Franco regime of an equally distributed guilt, alleging that "it was better to avoid talking about 'the good ones' and 'the bad ones'." Counterproposals often contained a generic condemnation of dictatorship, without explicitly mentioning the Franco regime. The PP's 1999 statement on the exile, for example, jumped directly from the civil war to the transition in its narration of recent Spanish history:

It is now 60 years since the end of the Civil War. . . . Our country, that had given a sad example of intolerance, resentment, and drive towards self-destruction, was able, 40 years after this sinister and bloody war, to show the world an example of tolerance, fraternity and the will to overcome in an exemplary democratic transition.[7]

The other typical feature of the PP's interventions was to underline the success of the transition, as if the traumatic memories would more easily be accepted if set against other, more positive recollections. The declaration of the PP from 2001 follows this pattern:

The Congress of Deputies condemns all dictatorships and authoritarian and totalitarian regimes of the past and the present, which are contrary to the fundamental rights and liberties of the individual . . . It [the Congress] recalls the historic success of our transition to democracy, explicitly founded on the will to reconciliation and on the overcoming of century-old conflicts that provoked the rupture of our cohabitation [*convivencia*] in 1936.[8]

This emphasis on the achievements of the transition, in fact, was the basis of the new historical master narrative of the successful transition, which was becoming increasingly dominant in those years. The ability to live in a democracy was experienced by the Spaniards as the negation of the myth of their ungovernable character. Apart from thus being based on the mythification of certain aspects of the transition, it also represents a democratic legitimacy based on popular satisfaction with the regime. This historical master narrative of the transition is essentially without a political stamp. The reference point is always the broad political consensus behind the most important decisions, and it is always emphasized, for example, that all the political parties from the communists on the left to the ex-Francoists on the right participated in the writing of the Constitution. This is a narrative of the coming of age of modern Spain by means of which Spaniards can celebrate Spain as a modern, European, and developed nation. Thereby the Spaniards are perceived to have finally overcome their history of decline with respect to earlier periods of splendor and their backwardness vis-à-vis European neighbors, which had dominated interpretations of Spanish history for over a century.

This focus on the transition, however, developed relatively late, and no broadly accepted occasion to celebrate the transition had evolved. Only the anniversary of the 1978 constitutional referendum, the Day of the Constitution, had been given an annual commemoration, but it clearly did not satisfy the perceived need to celebrate the transition. As a reaction to the relative commemorational vacuum around the historical master narrative of the transition, all the different twentieth and twenty-fifth anniversaries related to the transition to democracy invariably have experienced a commemorative resurgence since the mid-1990s, including those that had never been celebrated before, like the anniversary of the first elections. In accordance with the new historical master narrative, pride in the Spanish nation and in being Spanish is increasingly identified with the successful transition to democracy. This development corresponds to a wider process in Spanish society, which is shown by the fact that these celebrations generally represent relatively cohesive moments in which the Left and the Right, as well as Spanish nationalists and peripheral nationalists, can celebrate together.

When the PP came to power in 1996, this pride in the transition began to be turned into a useful token of Spanish national pride in the conservative party's search for ways of legitimizing Spanish nationalism. Mythifying the transition and its principal ingredients such as the Constitution, paradoxically, also meant that it became increasingly untouchable and inimitable. Elevating the Constitution to this kind of symbolic status thus preempted any kind of constitutional reform, which was rejected as an expression of nineteenth-century political instability. The latest development in this evolution was the appropriation by the PP of the concept of constitutional patriotism, originally proposed in the German context by political philosopher Jürgen Habermas, as a synonym for political-civic nationalism. Thus, the conference paper "Constitutional patriotism in the 21st century," which had been commissioned by the party leadership for the XIV party conference in January 2002, stated: "Spain is a great country, a nation shaped over the centuries. . . . A plural nation with a non-ethnically-based identity, but politically, historically, and culturally-based."[9] The emphasis is not only on the Constitution, but also on the fatherland to which that Constitution belongs, as well as on Spanish symbols, history, and culture. The reasoning departed from the fact that the transition to democracy is a source of pride in Spain today, and since the Constitution was generally understood as the expression of the transition, it could be turned into a legitimate object of pride.

The concept of constitutional patriotism provided a convenient way for the PP to express a national pride that was something other than nationalism, which was identified with the dangerous exclusionist practice of certain peripheral nationalists. "We are not nationalists," the program emphasized, although "we assume the idea of Spain with naturalness and without historical complexes." The constitutional patriotism of the PP thus does not work without a patria, and the conservative elites clearly used the concept as a politically correct label for their updated Spanish nationalism. The Constitution in this version is not first and foremost an expression of a series of civic and democratic values, but rather is seen as the essence of Spanish history and culture and a tangible proof of how successfully Spain has adapted itself to the political demands of present times: "We are living in a country, we are a country, that has been capable of doing these things [the transition]. And a nation which is capable of doing these things should confide in itself."

This new focus on the transition as a historical point of reference ran parallel to a reevaluation of the relationship between the present and the recent past. In February 2002, the PP thus finally agreed to explicitly condemn the Francoist dictatorship in an amendment to the original proposal on the adoption of measures of moral and economic restitution for those who suffered imprisonment and repression for political motives during the Franco regime. This was the first time that the Popular Party explicitly mentioned the Francoist dictatorship. The amendment was, however, rejected by the original proponents.

A common position on the recent traumatic past, however, was reached in November 2002. The parliamentary opposition had presented five different proposals, all aimed at making the government morally recognize the losers of the civil war, economically help the exiles and other victims, and officially support the reopening of mass graves that was beginning to take place all over Spain in those days. The date chosen to make this historical accord among the main Spanish political forces was hardly coincidental. The debate took place on November 20, 2002, the twenty-seventh anniversary of the death of Franco and the sixty-sixth of the death of José Antonio Primo de Rivera, founder of the Spanish fascist party, *Falange Española*. This time, the PP changed strategy and negotiated a common declaration with the opposition. From its early policy of silence on the recent past, via the various parliamentary initiatives of the opposition through which it developed a more active position, the conservative PP arrived at making propositions to the opposition. The most remarkable aspect of the accord, a compromise amendment to the five original proposals, was the explicit condemnation of the Francoist dictatorship and the civil war, and the expression of moral recognition to the victims of both. The agreement was preceded by a long introductory text that lauded the transition and explained its few shortcomings, which the declaration was aimed at repairing, and as such it followed the scheme of the earlier amendments by the PP. The governing party, however, this time went further than it had been willing to do before and recognized a fundamental difference between the two parts of the conflict, admitting that the nation had been split.

This acknowledgment of a divided nation was used to revalue the spirit of consensus and moderation of the transition, which was seen to have healed the wounds of the national community. The counterpart to this new vision of the history of the civil war and the Franco period was thus a further exaltation of the qualities of the transition.

This may be in harmony with how the Spaniards feel about the transition, but at the same time it may not exactly further the possibilities of a historical analysis of that period. Furthermore, the accord set the limits upon future dialogue with the past, that is, of historical investigation and of restitution to victims of history, by setting the condition that it should not be used to "reopen old wounds or add fuel to the flames of the civil conflict."[10] The danger that wounds might reopen was thus seen to persist; the myth of the ungovernable character of Spaniards is thus kept alive for the purpose of urging a certain sort of political and civic behavior.

The intention of the PP, and its reason for agreeing to negotiate a common position, was that the declaration should be understood to be definitive, which was clearly indicated in the debate:

> The accord is a point of convergence between all the parliamentary groups of this Chamber, today and forever. The idea is to get these questions out of the political debate . . . [A]ll the groups today want to put a credible end . . . to the rosary of parliamentary initiatives that have been made regarding this issue in the Chamber.[11]

The PP clearly wanted to remove the issue of historical justice in its various aspects from the political debate, once and for all, that is, to block the use of the past as a political weapon that had characterized its period in power. This had been the principal aim of the pact not to instrumentalize the past politically, and in this sense the new agreement represented a formal renewal of the old informal pact based on an updating of the historical status quo on which it was based.

This attempt to settle the political costs of the civil war and the Franco regime once and for all bore certain parallels to the so-called historians' dispute [*Historikerstreit*] in the mid-1980s in West Germany. The central issue of that discussion was the content of a West German national identity or self-perception. Despite the denomination, it was not an internal discussion among historians only, and since it was conducted mainly in national journals and newspapers it reached a broad public that became aware of the unsettled relationship with the past. From neoconservative revisionist historians came an attempt to relativize the Nazi crimes that made national identification and pride impossible. The revisionists were not particularly successful in their attempt to close the debate on the traumatic past of Germany. In fact, the Nazi past has reappeared frequently since, both in public debates and internal disputes among historians.

The principal parallel between the German and Spanish disputes was that, in both cases, conservative forces were trying to get around the problems of national identification created by the traumatic past. They therefore attempted to normalize the national past and settle the related political costs once and for all. In contrast to what happened in Germany, where politicians remained on the margin of the *Historikerstreit*, which did not involve official statements, in Spain the new pact was the result of a negotiation among political elites. It was thus, in the first place, a political rather than a public discussion, and it was sealed with an accord, in contrast to the result of the German dispute. The PP admitted that the blame for the civil war and the dictatorship was not evenly distributed, and that this period therefore had created victims to whom historical justice had been denied by silencing their suffering in official discourse. As a correlation, the PP expected the other political parties to stop the "rosary" of initiatives in the bit-by-bit rectification of historical justice. That is, that they would change the political use of the past that they had been adopting since the beginning of the PP government to come back to the modes of the old informal pact. The desire was to perpetuate the absence of a political use of the past conceived as an open process of permanent dialogue between the past and the present with this new formal accord. Instead of accepting the constant necessity to arrive at compromises regarding the national past, the government thus attempted to impede the presence of a particular past within political life. The accord is too close in time to draw any conclusions regarding its durability. It remains to be seen whether the past will now cease to be instrumentalized politically.

Notes

Parts of this chapter have previously been published in Spanish as "Usos políticos del pasado reciente durante los años de gobierno del PP," *Historia del presente* 3 (2004). All quotations have been translated by the author.

1. Juan G. Ibáñez, " 'Eran más los riesgos que las dificultades.' Diputados elegidos en 1977 y que continúan en el Congreso hacen balance," *El País Digital*, June 15, 2002.
2. "Declaración del Gobierno: 'Una guerra fratricida no es un acontecimiento conmemorable,'" *Ya*, July 19, 1986, 3.
3. José Lera, "Trillo dice de los brigadistas que 'más vale no resucitar hechos tristes,' " *El País*, November 11, 1996, 21.
4. Statement by deputy Alcaraz Masats from the IU, *Diario de Sesiones del Congreso de los Diputados*, hereafter *DSC*, 59 (2001), 2820.
5. Alcaraz Masats, *DSC*, 82 (2001), 4147.
6. Statement by deputy Silva Sánchez from the Catalan parliamentary group, *Diario de Sesiones del Congreso de los Diputados. Comisiones*, hereafter *DSCC*, 272 (2001), 8234.
7. Alternative proposition by the PP, *Boletín Oficial de las Cortes Generales*, hereafter *BOCG*, D-447 (1999), 13–14.
8. Amendment by the PP to the original proposal, *BOCG*, D-135 (2001), 4–5.
9. All citations are from Josep Piqué and María San Gil, "El patriotismo constitucional del siglo XXI," conference paper for the XIV National Congress of the PP (January 2002), http://www.pp.es, 4–5.
10. Compromise amendment, *BOCG*, D-448 (2002), 12–14.
11. Deputy Atencia Robledo from the PP, *DSCC*, 625 (2002), 20517.

Suggestions for Further Reading

Aguilar, Paloma, "Justice, Politics and Memory in the Spanish Transition," in Alexandra Barahona de Brito, Carmen González-Enríquez, and Paloma Aguilar, eds., *The Politics of Memory: Transitional Justice in Democratizing Societies* (Oxford: Oxford University Press, 2001), 92–118.

Aguilar, Paloma, *Memory and Amnesia: The Role of the Spanish Civil War in the Transition to Democracy* (New York: Berghahn, 2002).

Aguilar, Paloma and Humlebæk, Carsten, "Collective Memory and National Identity in the Spanish Democracy: The Legacies of Francoism and the Civil War," *History and Memory* 14: 1/2 (2002), 121–165.

Llobera, Josep R., "The Role of Commemorations in (Ethno)Nation-Building: The Case of Catalonia," in Clare Mar-Molinero and Angel Smith, eds., *Nationalism and the Nation in the Iberian Peninsula: Competing and Conflicting Identities* (Oxford: Berg, 1996), 191–206.

Núñez, Xosé-Manoel, "From National-Catholic Nostalgia to 'Constitutional Patriotism': Conservative Spanish Nationalism since the early 1990s," in Sebastian Balfour, ed., *The Politics of Contemporary Spain* (London: Routledge, 2004).

Serrano, Carlos, *El nacimiento de Carmen. Símbolos, mitos y nación* (Madrid: Taurus, 1999).

PART 2

New Nations

CHAPTER 5

Constructing Primordialism in Armenia and Kazakhstan: Old Histories for New Nations

Ronald Grigor Suny

Responding to an invitation to a conference in Armenia, I returned to its capital, Erevan, in July 1997 after a seven-year absence—a time in which Soviet Armenia had become the independent Republic of Armenia. The world I entered was not the one I thought I knew but one that had changed significantly. Armenians had gone through a decade of devastation, beginning with the struggle over Karabakh, an Armenian-populated region in the neighboring republic of Azerbaijan. This was followed by a destructive earthquake that killed twenty-five thousand and from which the country had never fully recovered, war with Azerbaijan with tens of thousands of victims, economic blockade, the collapse of the old Soviet economy and political order, and the creation of a new political system riddled with corruption, cronyism, and cynicism. Optimists spoke of a "transition to democracy" and the foundation of a market economy, but ordinary Armenian citizens experienced rapid impoverishment, radical social polarization, and dismal prospects for the future. Hundreds of thousands voted with their feet and left the country for Russia, Europe, or Los Angeles. In place of the tattered and discredited Soviet ideology, many in the political and intellectual elite espoused a fervent and increasingly intolerant nationalism.

A desperate loyalty to ethnicity and an unalterable sense of nation dominated the proceedings of the conference, which was held at the American University of Armenia. The conference organizers had invited me to speak about "prospects for regional integration" in the south Caucasus, a truly utopian topic at that moment of ethnic conflict among Armenians and Azerbaijanis, Georgians, Abkhazians, and Osetins. Reviewing briefly the nationalist reconceptualization of the Armenians in the nineteenth century from a primarily ethnoreligious to a secular national community, I discussed how the tiny Armenian state had become ever more ethnically homogeneous and nationally conscious during the Soviet period and raised the question: how can Armenians (or Georgians and Azerbaijanis for that matter) reconcile the idea of relatively homogeneous nation-states with the realities of trans-Caucasian politics and demography, which were formed by centuries of multinational empire and migration? Among ethnonationalists in south Caucasia the discourse of the nation—the notion that political legitimacy flowed upward from the people constituted as a nation—had narrowed to the view that the people must be ethnically, culturally, perhaps racially singular. The result has been ethnic cleansing and killing, deportations and forced migrations, and a series of enduring conflicts in Karabakh (between Armenians and Azerbaijanis), Abkhazia (between Georgians and Abkhazians), and South Osetia (between Georgians and Osetians).

My cursory survey of the three-millennium history of the region emphasized the long constitution of a shared Caucasian culture, a polyglot, migrating population, cities inhabited by diverse peoples, and soft, blurred, shifting boundaries between ethnic and religious groups. As examples of what I meant, I mentioned that "Baku and Tbilisi [the current capitals respectively of Azerbaijan and Georgia] were models of interethnic cohabitation; Tbilisi at one time had an Armenian majority, and Erevan was primarily a Muslim town at several points in its long history." The thrust of the talk was to question the usefulness of ethnonationalism in the current situation by proposing a more constructivist understanding of nationness in place of the primordialist convictions of the nationalists.

> At this point the positive effects of anti-imperialist nationalism metamorphose—one is tempted to say metastasize—into the negative effects of exclusivist, even expansionist, ethno-territorial nationalism. At this point, something else is needed—a revival of the more cosmopolitan pan-Caucasian tendencies of the past . . .[1]

The reaction to the talk was explosive. Leaflets were distributed the next day to all participants, pointedly challenging the assertion that Erevan had had a Muslim majority; newspapers and radio broadcasts attacked the speech. Hostile questions were directed to me at the conference, along with accusations that I was an "agent of the oil companies" and shared a secret agenda with the State Department! (After years of being suspected of being part of "the international Communist conspiracy" or, from the Soviet side, accused as a "bourgeois falsifier," I did not know whether to be relieved or embarrassed.) An angry crowd surrounded me as I was leaving the hall, shouting that I was *davejan* (a "traitor" in Armenian). I shouted back that I was a scholar and an Armenian, only to be told that I was no scholar and no Armenian. Security guards took me away to avoid further trouble. Personal attacks continued in the press, and a year later a book appeared in Erevan bitterly denouncing Western scholarship on Armenia, particularly my own work.

In Erevan that summer two fundamentally different languages of analysis met in a moment of mutual incomprehension. The clashing views of the nation evident in that confrontation have deeply affected, indeed distorted, scholarship on Armenia (as well as Georgia, Azerbaijan, and other former Soviet republics). In this essay I would like to explore the tension between the subversive investigation by scholars of the historical formation of ethnic, cultural, and national identities, on the one hand, and the actual practice of identity entrepreneurs, constructing (and simultaneously denying the constructedness of) identities, on the other. In some ways that tension can be expressed as the fundamental and apparently contradictory difference in the way the term identity is employed by academic analysts and in ordinary speech. Identity is both a category of intellectual analysis and of practice, that is, a category of "everyday social experience, developed and displayed by ordinary social actors, as distinguished from the experience-distant categories used by social analysts."[2] The analytical use of the term involves a recognition of the fragmented and contested process that goes into self or group identification, whereas the more common, everyday use of the term in normal "identity talk" usually defaults to an essentialist, often primordialist, naturalized language about a stable core, an actual unity and internal harmony. In this essay I explore the ways in which nation and national identity are reified, made into something real, that, while infinitely contestable, is no longer permitted to be contested in the public arena—at least not in certain proscribed ways. Those who question what has been set up as the "national" are either excluded from the national community—"you are no Armenian"—or punished, disciplined, and brought into line.

When people talk about identity their language excludes a sense of historical construction or provisionality and instead almost always accepts the present identity as fixed, singular, bounded, internally harmonious, distinct from others at its boundaries, and marked by historical longevity, if not rooted in nature. This loss of a processual sense of identification taking place over time is particularly acute in the rhetoric about national identity, which has become the universal category for modern political communities marked by an purportedly shared culture. Modern nations may be defined as those political communities made up of people who believe they share characteristics (perhaps origins, values, historical experiences, language, territory, or any of many other elements) that give them the right to self-determination—perhaps control of a piece of the earth's real estate (their homeland), even statehood and the benefits that follow. Like other identifications, they can be thought of as arenas in which people dispute who they are, argue about boundaries, who is in or out of the group, where the "homeland" begins and ends, what the "true" history of the nation is, what is "authentic" about being national and what is to be rejected. Nations are articulated through the stories people tell about themselves. The narrative is most often a tale of origins and continuity, often of sacrifice and martyrdom, but also of glory and heroism.

The post-Soviet states present a veritable laboratory of modern national identity formation. Comparison between republics, as well as intensive investigation of single cases demonstrate the ways in which identification is a multiple process that involves, first, the historical social positions (fluid, shifting, and discursively constituted as they may be) in which people find themselves, which shape, influence, and limit the possibilities of identification with some "others" and not with other "others." A young woman born in Stockholm, of parents who speak Swedish and identify themselves and her as Swedish, educated in Sweden, is more likely to identify herself as Swedish than as American, until, years later, she marries an American, migrates to Ann Arbor, and raises children born and educated in the United States. Proximity, distance, and length of time are key influences on stable and lasting associations and networks, whether kinship, friendship, collegial, or national and have powerful determining effects on identification with groups, location, and nation.

But a woman born in Tbilisi during Soviet times, of parents who speak Georgian and identify themselves and her as Georgian, educated in Georgia, may be more likely to identify herself as Georgian even after she marries a Russian, moves to Russia, and raises children born and educated in Russia. Her ethnic identity as Georgian remains fixed on her internal Soviet passport, and in a multinational state in which ethnicity provided both opportunities for social mobility (within the Georgian republic in this case) or serious disadvantages, that identity would remain potent. The Soviet example illustrates a second influence on identification when identity categories are externally generated, ascribed, or imposed by state or other authorities. In the Russian Empire and the Soviet Union state practices fixed subjects and citizens into legal categories—*sosloviia* (estates) and religious and ethnic designations in tsarist times; class and nationality categories in Soviet times—that gave them both privileges in some cases and disadvantages in others. Postcolonial studies in particular have contributed enormously to our understanding of how mapping, naming, census categories, statistical enumeration, and other practices of the modern state have delineated and fixed the more fluid distinctions generated by people, turning blurry differences into more visible, seemingly unalterable differences.[3] For the post-Soviet states the Soviet experience, for all the efforts to eradicate it, has been an indelible influence. The practice of fixing nationality in each citizen's internal passport on the basis of parentage rendered an inherently liquid identity into a solid commitment to a single ethnocultural group. Young people with parents who had different national designations on their passports were forced to choose one or the other nationality, which then became a claim to inclusion or an invitation to exclusion in a given republic.

Some theorists are already asking (as probably some of the readers of this essay are as well): why bother about identity? Why indulge in so much theorizing about such an abstract and contested term? The payoff of employing the concept of identity is threefold. Sensitivity to the fluidity of identities, as well as the naturalizing tendencies of identity talk, helps the researcher avoid, first, essentialism and, second, reification. Essentialism may be defined as the attribution of behavior or thinking to the intrinsic, fundamental nature of a person, collectivity, or state. Identity theory proposes an alternative to essentialist models of people or social groups by claiming that rather than having a single, given, relatively stable identity, persons and groups have multiple, fluid, situational identities that are produced in intersubjective understandings. Reification "is the apprehension of the products of human activity *as if* they were something else than human products—such as facts of nature, results of cosmic laws, or manifestations of divine will. Reification implies that man is capable of forgetting his own authorship of the human world,"[4] Identity theory instead emphasizes the historical and contextual generation of both categories and their effects. In this approach human agency remains central to the production of identities.

Although individual senses of the self may differ radically from one society or culture to another, it is possible to assert that there cannot be a group that does not possess some sense of shared commonality with others, even if it is just being in a certain room at ten past twelve, and a sense of difference from others in another room or with no room of their own. Cohesion of a group may depend on the particular articulation of the sense of commonality, and here a sense of shared past experience, that is, history, becomes important as a record of what binds the group together and distinguishes it from others. Nations are particular forms of collectivity that are constituted by a process of creating histories. Just as there are few groups without a sense of continuity, so there can be no nation without a sense of its own history. History contributes in several significant ways. Like the genealogies of ancient and medieval kings, history provides ancestry that legitimizes present-day loyalties. The art of "seizing and recording one's own history," writes Natalie Zemon Davis, not only contributes "a deepened sense of identity" but "an affective-political gain in enablement."[5] The national history is one of continuity, antiquity of origins, heroism and past greatness, martyrdom and sacrifice, victimization, and overcoming of trauma. It is a story of the empowerment of the people, the realization of the ideals of popular sovereignty. Whereas in some cases national history is seen as development toward realization, in others it is imagined as decline and degeneration away from proper development. In either case an interpretation of history with a proper trajectory is implied.

Beyond the specific narratives of particular nations is the metanarrative or discourse of the nation, the cluster of ideas and understandings that came to surround the signifier "nation" in modern times (roughly post-1750). This available universe of meanings allowed for the power of nations and nationalism to constitute collective loyalties, legitimize governments, mobilize and inspire people to fight, kill, and die for their country. This cluster of ideas includes the conviction that humanity is naturally divided into separate and distinct nationalities or nations. Members of a nation reach full freedom and fulfillment of their essence by developing their national identity and culture, and their identity with the nation is superior to all other forms of identity—class, gender, individual, familial, tribal, regional, imperial, dynastic, religious, or state patriotic. Though the nation may be divided or gradated along several axes, it is politically and civilly (under the law) made up of equals. All national members share common origins, historical experiences, interests, and culture, which may include language and religion, and have an equal share in the nation. The discourse of the nation both acknowledges that each nation is unique, with its own separate past, present, and destiny, yet recognizes the developmental process that gives every nation the conviction that the nation is always present, though often concealed, to be realized fully over time in a world of states in which the highest form is a world of nation-states. The national may be *in* people unconsciously and may need to be brought forth or willed into consciousness, but in this discourse the nation is never completely subjective but always has a base in the real world.

With the collapse of the Soviet Union and the emergence of dozens of new states, pundits, journalists, and often scholars made the simple assumption that coherent nations already existed, at least prefigured in the republics of the Soviet Union or federal Yugoslavia or Czechoslovakia. The prevailing narrative in the Soviet case, taken up by local nationalists, was that these nations had existed prior to the imperial conquest by the Bolsheviks, that they had been suppressed and denied their national expression during the long dark years of Soviet rule, and that they represented a population yearning for freedom, democracy, and capitalism. Left out of this narrative were the powerful effects on nation-making, rather than nation-destroying, of Soviet policies.

Soviet Roots, Post-Soviet Plants

In the last decade of the cold war, scholarship on Soviet nationalities shifted from a dominant view that the USSR was primarily a "prisonhouse of nations" in which national characteristics were being eroded by repressive and Russifying programs to a new paradigm that emphasizes the constructive formation of new national identities and the social consolidation of nations in many republics that occurred despite the more assimilationist, antinationalist, and often brutal policies of the Soviet regime. Recent scholarship has uncovered a complex process of nation-making that occurred as the ironic result of Soviet nationality and modernization policies and thwarted the Leninists' goal of a post-nationalist amalgamation of the peoples of the federation.[6] The picture is not a neat one. While some policies led to assimilation of smaller peoples, particularly in the Russian Federation, in many of the Union republics the titular nationalities became demographically more consolidated, better positioned in the intelligentsia and administrative apparatus, and more expressive in their national idiom. While most of the larger nationalities identified with their home republic, which effectively became territorialized nation-states (though without full political sovereignty), hundreds of thousands of Soviet people migrated from their original "homelands" to become dispersed throughout the vast Union that they considered their extended homeland.

Before the revolution most peoples of the Russian Empire were only beginning to develop a national self-consciousness, and then largely among their elites. In south Caucasia or central Asia identity was primarily with coreligionists, fellow speakers of one's language, and regional "civilizations," rather than with a fixed and bounded homeland. Soviet nationality policy, based as it was on national territorial autonomy and *korenizatsiia*, the "rooting" of national culture and cadres in the national areas, enhanced a sense of national homeland, and the modernization program that promoted rural to urban migration contributed to "the more rapid nationalization of the masses."[7] In the 1920s Soviet officials attempted to draw the boundaries of administrative units as close as possible to the apparent boundaries of ethnic communities. But since ethnicity was an inherently fluid identity and lines between groups were often blurred, officials and ethnographers had to make decisions about who belonged where. The aim, however, was to have ethnicity, territory, and political administration correspond as clearly as the science of the day enabled. Through the course of Soviet history boundaries were changed to conform to new understandings of national distinctions, but the basic principle of territorializing ethnicity and linking both to politics remained constant. Even after Joseph Stalin, Soviet leader from 1922 to 1953, shifted the Russophobic emphasis of early Leninist nationality policy toward promotion of the Russian language and culture in the early 1930s, the regime continued to support the ethnic nationalization of a reduced number of the larger republics. The Caucasian republics, for example, became over time increasingly homogeneous, and in the last decades of Soviet power Russians as well gradually began to migrate out of the region.

Committed as it was to the international equalization of its peoples and raising the more backward to positions equal to the most advanced, the USSR engaged in what has been referred to as affirmative-action programs designed to advantage the indigenes in their own national territories.[8] In the early Soviet years affirmative-action programs aided non-Russians to achieve native language education, to advance socially, and gradually to occupy positions of power in industry, education, culture, the party, and the state. But later, in the post-Stalin period, affirmative action in a context in which the "disadvantaged" nationality to be advanced now was in fact politically advantaged, even entrenched, in its own republic gave the titular nationality a double advantage—both in access to education and jobs, and as the principal distributors of advantages. Such programs only reinforced the sense of non-titulars, like "Europeans" in central Asia or Armenians in Azerbaijan, that their ethnicity was a positive mark of discrimination. An Armenian KGB officer in Georgia remarked to me in the 1970s that everyone understood that "this was a Georgian shop." There was a widespread sense among Armenians in Georgia that they would occupy subordinate positions, do most of the real work, while the Georgians on top would receive most of the prestige and privileges.

Instead of equality two kinds of hierarchy developed in the USSR: an imperial relationship between the Soviet center and the non-Russian peoples, in which the increasingly territorialized nations remained subordinate to the dictates and requirements of Moscow's all-Union goals; and a "national hierarchisation," in which certain nationalities, like the titular nationalities of the republics, were considered superior to others within the republic, and Russians often held a special place of privilege no matter where they lived.[9] From the earliest years of the Soviet state the Bolsheviks spoke of "backward" and "civilized" nations, "peasant" and "proletarian" peoples, and the Russians were among the more civilized and proletarian. The state categorized ethnicities by size and development—*natsiia* (nation), *natsional'nost* (nationality), *narod* (people), *narodnost'* (small or less developed people), and *plemia* (tribe)—implying that some were superior to others. Hierarchy was reinforced in most republics as the titular nation or nationality developed a sense that it possessed that republic and that other ethnicities, except perhaps the Russians, were not entitled to the same advantages. Such policies were particularly egregious in Georgia, where Abkhazians and Osetins experienced discriminatory treatment, and in Azerbaijan, where the Armenians of Karabakh protested against restrictions on their language and culture and repeatedly petitioned for merger with the Armenian republic next door. Union-republic nations had more advantages than nationalities whose homeland was merely an autonomous republic, and peoples without territories of their own fared worst of all. Even as non-Russians experienced upward social mobility, the very linking of ethnicity and various social benefits—admission to university, advancement in the workplace—created resentment that would later be exploited by nationalists.

In contrast to the expectations of both Marxism and the modernization theory—that industrialization and urbanization in either its capitalist or socialist variant would lead to an end to nationality differences and conflicts—not only was nationality preserved in the Soviet Union, but also the power and cohesion of nationalities and their elites were enhanced. The achievement of greater (though hardly complete) equality among nationalities did not lead to the "withering away" of interethnic hostilities. Rather social mobilization intensified interethnic competition for limited social resources, while urbanization and education led to "heightened national self-consciousness and increasing national separatism among the more socially mobilized members of each national community."[10] Russification occurred, both spontaneously and through government programs, but in the Union republics in particular indigenous intellectuals defended and promoted their own culture and language. Powerful national elites emerged in the late Soviet period, as Nikita Khrushchev (1953–1964) and particularly Leonid Brezhnev (1964–1982) permitted national communists to remain in power for many years. *Zemliaki* (people from thes same ethnicity or region) networks were particularly tight in the Caucasus and central Asia, where local traditions emphasized loyalty to kin, clan, region, and close friends. The highly centralized command system of the Stalin years loosened its grip on the national republics, and by the last decades of Soviet power nationalities experienced an unprecedented degree of local autonomy.

Nationality was institutionalized into the Soviet system as a category of identity, a passport to privilege (or discrimination), and a claim to political power in national republics. Moreover, the idea of nationness fluctuated between a more contingent understanding of nationality as the product of historical development to a more primordial sense that nationality was deeply rooted in the culture, experience, mentality, and even biology of individuals. Soviet theorists held contradictory views: that national differences would eventually grow less distinct and Soviet peoples would meld into a single Soviet people; and that nationality was passed on, like genetic traits, from one generation to another. Even as class evaporated as an official status in Soviet life, nationality became ever more primordial. At the end of the 1930s the Soviet authorities celebrated the putative "anniversaries" of the epics of various Soviet peoples: the Georgian *Vepkhistqaosani* (*Knight in the Panther's Skin*) by Shota Rustaveli (1937), the Russian *Slovo o polku Igoreve* (*Lay of the Host of Igor*) (1938), the Armenian *Sassuntsi David* (1939), and the Kalmyk *Jangar* (1940). An industry of ethnographers and ethnologists developed an enormous body of theory in the post–World War II years elaborating the ancient roots and ethnogenesis of Soviet peoples. The famous "fifth point" in the Soviet internal passport, which listed the holder's nationality, was based on parentage.[11] "Every Soviet citizen was born into a certain nationality, took it to day care and through high school, had it officially confirmed at the age of sixteen and then carried it to the grave through thousands of application forms, certificates, questionnaires and reception desks. It made a difference in school admissions and it could be crucial in employment, promotions and draft assignments."[12]

With the political openings offered by the reformer Mikhail Gorbachev, autonomous political movements emerged in the Soviet Union, and in national republics they quickly became the vehicles of nationalist expression. Ecological, politically democratic, and nationalist activists, as well as "liberal" communists, took advantage of Gorbachev's policies of *glasnost'* (openness) and *perestroika* (restructuring) to push for greater public participation in decisionmaking. The progressive weakening of the central Soviet state and the Communist Party of the Soviet Union opened the way for three distinct political patterns in the non-Russian republics. First, in a few republics—Armenia, Estonia, Georgia, and Latvia (and in Chechnya and Tuva within the Russian Republic)—noncommunist nationalist leaders took power with broad support of the population. Second, in a number of republics—Kazakhstan, Kyrgyzstan, Lithuania, Moldova, Russia, and Ukraine—former communists quickly adapted their political agendas to fit the new postcommunist period of nation-building and to varying degrees adopted programs of democratization and marketization.

Third, old communist elites—in Azerbaijan, Belarus, Tajikistan, Turkmenistan, and Uzbekistan—stubbornly attempted to hold on to power, thwarted the aspirations of nationalists, and threw up a façade of democracy and nation-building while essentially maintaining a Soviet-style distribution of power.

These patterns were quite unstable, however, and republics shifted from one to another. In general democratic institutions and practices gave way to more authoritarian ones in Armenia, Azerbaijan, Belarus, Kyrgyzstan, Kazakhstan, and Tajikistan. Civic nationalism tended to be undermined by ethnocultural nationalizing in the newly independent states. In the absence of powerful constituencies favoring Western-style capitalist democracy, a furious search for an "authentic" national identity and politics occupied both state officials and the cultural intelligentsia. Although difficult to measure, popular adherence to a national identity appears to have strengthened over time, while identity with the old Soviet Union has declined. Particularly strong in Armenia and Georgia, national identity competed less well with local identities or supranational Islamic (non-European) identifications in Azerbaijan and central Asia.[13] The Soviet practice of ascribing ethnonational identities at the republic level had powerful popular resonance, but older patterns of clan, tribe, and regional identification undermined effective commitment to the nation in several republics, most notably Azerbaijan and Tajikistan.[14] The overriding identity with the Soviet Union, deeply ingrained in Russians in particular, gradually evaporated in the course of the 1990s, although not without regret and even resistance among the older, more conservative generation.

To illustrate the struggles over constructing national identities in the post-Soviet period, I shall look at two polar cases—one in which national identity was largely a Soviet product and where linguists and historians are actively "recovering" and consciously constructing identities (Kazakhstan); and a second with an unusually strong primordial identity and a fierce opposition to notions of constructivism (Armenia).

Kazakhstan

When in 1992 political scientist Bhavna Dave asked her Kazakh informants about the "plight" of the Kazakhs, how they, their language, and culture had "become marginalized" in their own homeland, she heard consistent responses: "it was the Soviet system, its unmitigated policy of Russification and colonization, the 'genocide' [the loss of about two million Kazakhs during the forced sedentarization of the nomadic Kazakhs in the early Stalin period], the influx of settlers to till the so-called Virgin Lands [in the late 1950s] that resulted in this unfortunate state of affairs."[15] In this late Soviet and post-Soviet construction of the recent past all agency passed from the Kazakhs to the "Soviet system," and Kazakhs were rendered victims of a brutal and alien state. The moment of independence just a year earlier had essentially jumpstarted a new era in Kazakh history that was starkly contrasted to the dark experience of the Soviet period. With statehood would come the revival of the national culture and the reversal of the Russification that had been imposed by the Soviet regime. Yet the experience of ordinary Kazakhs included more than memories of oppression and Russification. The modernizing project of the Soviet government had had profoundly transformative effects on the republic, many of which were judged positive by ordinary Kazakhs.

In contrast to other southern Soviet republics, where the national languages dominated over Russian, in Kazakhstan the Russian language was overwhelmingly the language of urban Kazakhs—not to mention the more than 50 percent of the population that was not Kazakh. Although the government and party apparatus had been effectively Kazakhized from the 1960s, the elite, as well as the great bulk of the educated population, preferred Russian to Kazakh both in their official and daily lives. Since the urban centers of Kazakhstan had largely been Russian, Kazakhs moving into towns quickly adapted to the dominant language. About 40 percent of Kazakhs could not express themselves in their "mother tongue" and some three-quarters of urban Kazakhs used Russian rather than Kazakh in everyday conversations. Kazakh had a low status among non-Kazakhs, and few bothered to learn the language. Russian was understood by Kazakhs to be the vehicle for social advancement.

At the same time the affirmative-action policies of the Soviet government promoted Kazakhs into positions of influence, gave them preferential treatment in admissions to higher education (which are almost entirely in Russian), turned them from nomads into urban settlers. Under the long tenure of Kazakh party chief Dinmukhammed Kunaev (1959–1962, 1964–1986), Kazakhs became the dominant nationality in the state and party, but political success required cultural competence in the ways of Soviet life, most importantly in the knowledge of Russian. Russification was rampant, and yet full assimilation did not occur. Distinctions between nationalities, both ascribed and experienced, remained, and the strong sense that ethnicity was deeply rooted in the human personality was reinforced in everyday Soviet life. "I like to speak in Russian," one of Dave's informants reported. "Yet I am a Kazakh at heart and will never think of myself as anything else. I love Abai as much as I love Pushkin, even though I have never read him in Kazakh."[16] Her primordial idea of Kazakhness contrasts vividly with the equally essentialist notion of those Kazakh nationalists who insisted "no language, no nation."

Before the Soviet period Kazakh collective identity had been based on its nomadic life. Indeed, the term "Kazakh" meant nomad, and Kazakhs (called Kyrgyz in tsarist times) distinguished themselves from other central Asians who lived a sedentary life.[17] Rather than with any notion of "nation," the nomads most strongly identified with their genealogical linkages, either the tribal confederation (*zhuz*) or smaller groups (*ru* or *taipa*). A Kazakh intelligentsia promoted literacy in Kazakh before the revolution, published a newspaper, *Qazaq*, that reached 8,000 subscribers, and in 1906 formed a patriotic organization, Alash, that came to prominence in the revolutionary years. But nationalism among the literati should not be equated with mass allegiance to an idea of the nation. The Soviet state's "nativization" programs of the 1920s and 1930s assisted the development of a standardized literary language that was employed in official institutions. Kazakh membership in the Kazakhstan Communist Party grew from 8 to 38 percent in four years (1924–1928), but these developments pale before the disaster of the late 1920s and early 1930s. Led by F.I. Goloshchekin, the party carried out a "small October" to transform the Kazakh way of life and eliminate traditional social relations. The state authorities ordered the collectivization of Kazakh herds and compelled the nomads to settle on the land. The herdsmen resisted by slaughtering their livestock. Hundreds of thousands fled to China, and in the chaos of collectivization over 40 percent of the Kazakh population was lost. The demographic catastrophe was later compounded by Kazakh losses in World War II and the influx of Slavic and other settlers in the late 1950s during the Virgin Lands campaign. Kazakhs became a minority in their own republic.

At the same time imperial modernization created a new Kazakh society: party and state officials, intellectuals with privileged access to state-subsidized institutions, and a working class tied to state industry. Upwardly mobile Kazakhs imbibed many of the values of Soviet modernization, even as they complained about the excesses of Stalinism and the failure of the system to meet its own standards of justice, equality, and material well-being. In the view of nationalists, modern Soviet Kazakhs resembled the *mankurts* of Chingiz Aitmatov's novel, *The Day Lasts a Hundred Years*, deracinated, denationalized amnesiacs without a sense of the past. Like other central Asians, Kazakhs did not participate in dissident or nationalist movements before 1989—the sole exception being the street protests of December 1986 against the installation of a Russian as head of the Kazakh party. By the time *glasnost'* and *perestroika* was opening up the "blank spots" of Kazakh history, the removal of Kunaev and his replacement by a Russian from outside the republic violated the deep feeling that Kazakhstan ought to be governed by the titular nationality.

In the discourse of the nation, culture is the source of political power. The right to rule belongs to the people conceived as a nation. A coherent, bounded, conscious nation must exist as the foundation of the state's legitimacy. Soviet state practices spent much time and energy establishing the original moment of ethnogenesis. Appearance of the ethnonym in travelers' accounts or other sources was often enough to conclude that a nation existed. For the Kazakhs it was eventually settled that the "nation" was formed in the mid-fifteenth century. Independence in 1991 radically changed the political salience of nationality and nationalism. Overnight a radical status reversal turned the ethnic Kazakhs from a subordinate people in a multinational empire into the "state-bearing" nation in a new state, while the former "elder brother," the "Russians" (actually Russian-speaking peoples) of Kazakhstan found themselves no longer living in their Soviet homeland but a beached diaspora in a new, potentially foreign state. The Communist Party chief, Nursultan Nazarbaev, easily converted to national leader, even as he resisted the call of independent nationalists for a more vigorous nationalizing program. He believed that Kazakhstan now required energetic state intervention in the cultural sphere, particularly in the development of the language of the titular nationality, to foster the consolidation of nationhood. In a major policy statement in the fall of 1993, he asked, ". . . to what can we turn if the previous [socialist] tenets have proven bankrupt?" And he answered, to cultural traditions, to one's historical cultural roots, which "enable a person to 'keep his bearings' and adapt his way of life to the impetuous changes of the modern world."[18] Kazakhness, Kazakh language, and traditions now took on a new value, one that contrasted markedly with the marginalization of Kazakh culture in late Soviet times.

At first it appeared that ethnicity and ethnonationalism would be the easy fallback position of state builders in the post-Soviet republics. Yet Kazakhstan, like a number of other post-Soviet states, experimented with an ethnically inflected variant of civic nationhood. In Kazakhstan the government maneuvered between the legacy of Soviet internationalism and an emerging ethnonationalism. Employing a kind of retreaded Soviet internationalism, former party boss Nazarbaev proposed a Eurasian identity for

Kazakhstan, linking Russians and Kazakhs into a single category. Kazakhstan was seen as a crossroad of civilizations, with legal protection for all peoples in a nonethnic state. "But, just as Soviet-era internationalism ultimately had a Russian face (holding a privileged position for ethnic Russians in the evolutionary march toward the 'bright future'), post-Soviet Kazakhstani state ideology had a Kazakh face, singling out Kazakhs for linguistic, demographic, political and cultural redress."[19]

In a rerun of the original *korenizatsiia* program of the 1920s, the state promoted Kazakh media, higher education in Kazakh, greater Kazakhization of the state apparatus, and repatriation of diaspora Kazakhs. Kazakh would be the state language, and Kazakh-language education would be stressed. On the "history front" scholars set out to find a more ancient lineage for the Kazakh state. "According to several informants in the Institute of History and Ethnography, the institute's director, a powerful ally of the president, issued an instruction to researchers to find the roots of Kazakh statehood in the Sak period (the first millennium BC). This was a clear departure from established historiography that located such statehood in the mid-fifteenth century."[20] One scholar attempted to incorporate Chinggis Khan and his empire into the Kazakh past in order to show that the Kazakh were "a more ancient and historically well-known people than the Mongols."[21]

The efforts of historians, as well as ethnographic expeditions sponsored by the state, aimed at ethnicizing the past of Kazakhstan, erasing its more multiethnic features, and establishing an ethnic Kazakh claim to territory. The experiences of pre-Kazakh Turkic tribes were assimilated into a Kazakh narrative.[22] The cultural activists found ancient heroes, called for preservation of monuments, and organized excavations. The Kazakh state was imagined as a caring, kind mother; Kazakhs were envisioned as a generous, hospitable people who opened their arms to other peoples; Kazakhstan, then, where Kazakhs were the first among equals, was a place where many nationalities could coexist. While the Kazakh national anthem proclaimed how the Kazakhs had suffered "on the anvil of fate, from hell itself," and the state emblem emphasized the antiquity and indigenousness of the ethnic Kazakhs, the successive drafts of the constitution (1993, 1995) moved in an internationalist direction. The preamble to the second constitution was boldly inclusive of all peoples of the republic. Its first sentence reads, "We, the people of Kazakhstan, united by a common historical fate . . ." A later article stated even more clearly: "No part of the people . . . can appropriate to themselves the sole right to exercise state power."[23] The winning design for the state flag was certainly symbolic of Kazakh ethnic dominance—a sky blue background, a golden sun, and a woven Kazakh design. But the sun could be understood as inclusive in a way that the Islamic crescents of the Azerbaijani, Turkmen, and Uzbek flags could not.[24]

While Kazakh nationalists expressed their anxiety about the loss of their language in the Kazakh-language press, the Russian-language media waxed nostalgic about the defunct Soviet Union. "Hankering for the unitary Soviet State is expressed openly," writes Pål Kolstø, "and indirectly one may infer that the editors do not accept the legitimacy of the Republic of Kazakhstan."[25] As Nazarbaev set out on the road of nation-making, he was faced, not only by Kazakh nationalists dedicated to Kazakh cultural dominance in the new state and the threat of massive Russian out-migration with the consequent loss of skilled labor, but also by a general indifference to the project of nationalizing the country. Dave found that in the early 1990s "most Kazakhs remain as apathetic to the nationalizing state as they were indifferent to the communist ideology. Soviet-style internationalism is in fact closer to their life experience than is the ongoing ethnicization of personal identities and the public sphere by a nationalizing state."[26]

Nazarbaev's nationality policy, pulled as it is between ethnic and civic conceptions of the nation, nevertheless allowed for stable and tolerant relations within the bicultural population of Kazakhstan. Priority was to be given to reviving Kazakh ethnic culture "because it cannot be sufficiently developed in a true sense in any other place than Kazakhstan."[27] While colonial victimization was to be redressed, the government supported the consolidation of both ethnic identities and supranational state identities in a multicultural setting. However, state builders and nation makers were not the only ones engaged in identity construction. Just as the state was promoting Eurasian, Kazakhstani, and Kazakh ethnic identities, a renewed pride in lineage identities (the *ru* and *zhuz*-based genealogies) emphasized subnational affiliations.[28] It remains unclear how in its search for nationhood Kazakhstan will be able to construct both supra-ethnic and subnational identities or how these identities will intersect, reinforce, or undermine one another.

Armenia

A people with a long written tradition (dating from the fifth century AD), with a past that includes numerous polities, dynasties, and continuous institutions (like the national church), Armenians enjoy a rich repertoire of symbols, legends, and historical accounts with which to construct a modern national consciousness. In sharp contrast to Kazakhstan, Armenia was the most ethnically homogeneous of the Soviet Union republics, with a high level of literacy in the Armenian language, and no real challenge to its ethnic dominance of its own republic. Armenians, however, were plagued by a sense of national danger. The republic was the smallest in the USSR in territory. High levels of migration from the republic, the loss of national sentiments among the diaspora, as well as the affinity for Russian-language education of much of the elite contributed to a presentiment that what a genocide early in the twentieth century had not accomplished might occur in a more gradual manner—a so-called white genocide through acculturation and assimilation. Through the modern period the historical territory of Armenia had been denuded of Armenians by successive Turkish governments (most fiercely in the genocidal massacres and deportations of 1915), and the existing state of Soviet Armenia represented only a tiny fraction of a once vast homeland. Not only were the lands now occupied by Turkey gone, but two formerly Armenian areas, Nakhichevan and Karabakh, were in the neighboring republic of Azerbaijan. Though Azerbaijanis were as secularized as Armenians after 70 years of Soviet power, many Armenians linked them as Turks and Muslims with the Anatolian Turks who had devastated historic Armenia. The sense of national danger apparent in the first public demonstrations in late 1987, aimed to close down a nuclear power plant and a synthetic rubber factory, exploded early the next year in a more militant political movement that called for unification of Karabakh with the Armenian republic. Demonstrations were met with a pogrom of Armenians in the Azerbaijani industrial town of Sumgait, and the anxiety about annihilation and genocide became palpable.

Among Armenians the themes that through repetition constitute the deep weave of tradition include the antiquity of the people, its indigenous and continuous occupation of the "homeland," the unique and significant role of Armenians in history (the first Christian nation, defenders of Christianity at the frontiers of Islam), and a constant struggle for survival and freedom. History is told as an epic, complete with heroes and martyrs, great sacrifices and persistence, treacherous enemies and unfaithful friends. As they tell their story, Armenians have been betrayed repeatedly, abandoned by great powers, invaded by uncivilized barbarians, and yet have survived. Often without a state of their own, Armenians have managed to remain constant to their ideals, thanks to the continuity of the national church. In one form or another these narrative elements can be found in the earliest Armenian texts—in Agathangelos, Eghishe, and Movses Khorenatsi. The narrative was then popularized, particularly in the nineteenth century, in poems, plays, novels, and spread through the periodic press and the burgeoning school system established by Armenians in the Ottoman and tsarist empires and in the diaspora. The clerical establishment was eventually forced to give way to a more radical, secular intelligentsia, the precursors of a revolutionary elite at the turn of the twentieth century. But history ruptured abruptly in 1915 (and again for many in 1920), first with the Ottoman genocide of Armenians and then with the Sovietization of the tiny Armenian republic. For most Armenians the recovery of an independent statehood in 1991 meant the revival of the nation, despite the catastrophic economic and social collapse experienced by independent Armenia.

The story of Soviet Armenia parallels, in interesting ways, the formation of the state of Israel: a part of the ancient "homeland" was reconstituted as a national state to which dispersed Armenians could return under the protection of a great power. Besides Soviet programs of "nativization" and the cultural nationalization of Armenia, the territory of the republic was demographically Armenized with the in-migration of Armenians and the sometimes involuntary deportation of Azerbaijanis. On several occasions Stalin's government moved traditionally Muslim peoples out of Armenia, in some cases exchanging populations with Armenians deported from Nakhichevan. Once the conflict over Karabakh became violent, in 1988–1989, hundreds of thousands of Azerbaijanis left Armenia for Azerbaijan. Migration of Armenians in the opposite direction accelerated after the January 1990 violence against Armenians in Baku, the capital of Azerbaijan. By the 1990s independent Armenia had become essentially a mono-ethnic state.

Political scientist Razmik Panossian has isolated a central romantic strand in Armenian nationalist expression that he sees running from the writer Levon Shant (1869–1951) through the emigré activist Edik Hovhannisian to post–Soviet Armenian theorists.[29] In this vision the Armenian nation is a historical constant, held together by blood, territory, religion, language, and history. As Shant put it, the individual cut off from the nation is like "a word outside a sentence; it has no role; and it has and does not have meaning. In order to receive a role and a certain meaning, to be able to express its real meaning and inner nuance, it must be woven into a sentence."[30] More mystically, Hovhannisian declares, "Not only the living, but also the dead speak in the national will. The past speaks, as well as the puzzling future."[31] Taking on the modernist, constructivist approach, Hamlet Gevorgian of the Armenian Academy of Sciences retorts,

What "re-creation" of historical memory is it possible to talk about in the case of a people who has continuously maintained and visited for sixteen centuries the memorial of the inventor of its alphabet, and whose main cathedral at Holy Ejmiatzin has been operating continuously for seventeen centuries . . .[32]

The antiquity and continuity of the Armenian essence is both a rejection of the denial of the reality of the nation repeated by both Marxists and modernists, as well as an implicit statement of the superiority of Armenian claims to territory and authenticity to those of more recently constructed "nations" like the Turks and Azerbaijanis.

In the post-Stalin years Soviet Armenian historians promoted an insistently national thematic: waging an effective guerrilla war against denationalization of their history. The story of the republic of Armenia was told as a story of ethnic Armenians, with the Azerbaijanis and Kurds largely left out, just as the histories of neighboring republics were reproduced as narratives of the titular nationalities.[33] Because the first "civilization" on the territory of the Soviet Union was considered to have been the Urartian, located in historic Armenia, the ancient roots of Armenian history were planted in the first millennium BC. Urartian sites and objects of material culture were featured prominently in museums, and the regime celebrated the 2,700th anniversary of the founding of Erevan (the Urartian Erebuni). A revisionist school of historians in the 1980s proposed that rather than migrants into the region, Armenians were the aboriginal inhabitants, identified with the region Hayasa in northern Armenia. For them Armenians have lived continuously on the Armenian plateau since the fourth millennium BC, and Urartu was an Armenian state. A rather esoteric controversy over ethnogenesis soon became a weapon in the cultural wars with Azerbaijan, as Azerbaijani scholars tried to establish a pre-Turkic (eleventh century) origin for their nation.[34]

The nationalist thrust of Soviet Armenian historiography extended into a fierce critique of foreign historians who attempted to question certain sacred assumptions in the canonical version of Armenian history.

A young historian in post–Soviet Armenia, Armen Aivazian, begins his critical review of American historiography on his country by declaring, "Armenian history is the inviolable strategic reserve [*pashar*] of Armenia."[35] His views, hailed by his countrymen, provide a window into the particular form of historical reconstruction of Armenian identity and historical imagination that dominates post–Soviet Armenian historiography.[36] His tone is militant and polemical, for his self-appointed task is to defend Armenia from its historiographical enemies. "From the point of view of Armenia's national (internal, civil, and foreign, international) security," he tells his readers, "in its consequences Western pseudo-Armenology is more harmful and dangerous than Turkish–Azerbaijani historiographical falsification because this is the real basis of the propaganda carried out on an international scale against the interests of Armenia and is also a constituent part of that propaganda."[37]

His focus in the first part of the book is on my collection of essays, *Looking Toward Ararat* (1993), in which "can be found the best expressions of the arguments for American 'Armenology's' anti-scientific and strongly politicized position and essence."[38] The argument of *Looking Toward Ararat* is that "Armenian essentialism has reinforced exclusiveness, ethnic isolation, and divisiveness within the [Armenian] community."[39] In its place I propose

> a more open understanding of nationality, one determined equally by historical experiences and traditions and by the subjective will to be a member of a nation. A distinction is drawn between a national essence or spirit, features that do not stand up to historical analysis, and a national tradition, a cluster of beliefs, practices, symbols, and shared values that have passed from generation to generation in constantly modified and reinterpreted forms.[40]

Reducing Armenianness to a "cluster of beliefs," and the like, is truly offensive to Aivazian, who puts forth a genetic theory of the Armenians. A people formed definitively in the sixth–fifth centuries BC, the Armenians share common genetic features that make them recognizable through time and around the globe. Although migrations and invasions have brought Armenians in contact with other peoples, he argues, their high rates of endogamy have preserved their essential biological features. Rather than distinguished primarily by culture or traditions, Armenians are biologically distinct. The primordial base of the nation is rooted in its genetic makeup, which is then reflected in its cultural production. Nation is not a choice but a given.

In sharp contrast to the primordializing direction of post–Soviet Armenian historiography, several studies by anthropologists have contributed a more dynamic picture of identity construction through time and space.[41] In a remarkable piece of research, based on extensive field work in Armenia during the worst period of material and spiritual devastation (1989–1994), historical anthropologist Stephanie Platz demonstrates that in the chaos of a collapsing economy, blockade by neighboring states, and the early stages of the war over Karabakh, Armenians found meaning and motive for their actions through their national identity, their reliance on family and kin ties, reliance on readings of historical experience, and a strong sense that authentic Armenian virtues would get them through the current difficulties. Even as social relations broke down under the strains of life without heat and light, a memory of a more authentic Armenia remained. Platz's Armenian informants repeatedly referred to the "time before" when Armenia was normal, people kind and hospitable to one another, when they had everything, when the country was disciplined and life guaranteed.[42] The nostalgia for the times lost, for a recent "golden age," was clearly a memory for an imagined, reconceived past but one that had been familiar, where life had been more predictable. But rather than stemming from uncontested fixed characteristics, Armenian identity was fraught with ambivalence and could be employed with positive and negative meanings. Contingent events like the earthquake of December 7, 1988 are "absorbed into a single historical narrative, which included massacres, genocide, environmental pollution, ethnic violence and state domination."[43] And even a marginal movement of UFO enthusiasts interpreted the arrival of extraterrestrials through the prism of Armenian national history. "Through the representation of history, Armenians resisted rupture and regression, by constructing a national space-time through social memory. In the face of adversity, *hayut'yun* [Armenianness] itself, propagated through discourse as a perpetual ideal, enabled Armenians to locate themselves in historical time and national space."[44]

For a people living in a republic that is nearly 100 percent Armenian the idea that national identity can be selected is far-fetched. Ethnic homogeneity within

Armenia precludes the kind of multiculturalist imaginary that affects the United States or Western Europe, or, indeed, the multinational empires of Armenia's past. But Armenians are a nation divided between those who live in the independent republic and those living in the diaspora, where conditions of choice, preservation, and acculturation are a daily matter. The Armenian genocide of 1915, in many ways one of the most potent sources of twentieth-century Armenian identity, appears to resonate far more loudly in the Armenian diaspora communities than in the republic itself and has become the perpetual sign of Armenian victimhood. Diaspora newspapers and journals constantly refer to the campaigns of the Turkish government and its supporters to deny that the events of 1915 qualify as genocide. The sense that Armenians could be extinguished as a people engages many of them in a continual effort to remind non-Armenians of the particular suffering of Armenians. Both in Armenia and the diaspora, histories are being constructed as part of the effort to give content to Armenian identity, yet at the same time most Armenians, and even their historians in the diaspora, resist the turn toward a constructivist view of history and identity.

Nationalists often strive to get history "right." In their "objectivist" reading of the past—showing the past as "it actually was"—they set themselves up as representing the only true account. This pretension to objectivity is also a pretension to an untroubled authenticity of a single reading. Reification of a historical reality is a powerful claim to the legitimacy of the nation and particular claims to territory and statehood. If the nation is real, ancient, and continuous, then in its own view (and in the discourse of the nation more generally) it is not vulnerable to counterclaims by other socially constituted groups, like classes, which, it claims, are not "real" in the same way. Constructivists, on the other hand, are more open to viewing historical writing as one possible account of the past. It is not that they consider all accounts as equally compelling but rather understand that professional historical writing abides by certain critical conventions—adherence to documentable facts, coherence, plausibility, neutrality. Such criteria can then be used to judge the value of competing interpretations.

In social science the very process of constituting a political community in the form of a nation has been given a positive valence. Political integration of localities or tribes into coherent nations was part of the project of modernization and was lauded by modernization theorists.[45] While nationalism (because of its affiliations with revolution and the Left) was suspect in the minds of many Western policymakers during the first great decolonization after World War II, political analysts were even more troubled by tribalism and social fragmentation than they were with efforts of nationalists to construct new, coherent communities on the model of Western nations. Democracies in particular require a clearly defined, bounded population that then has the right to be represented.[46] Nation is a convenient and powerful form of identification that speaks precisely to these conditions. If the irony of Soviet nationality development was that an antinationalist state helped create nations within it, the irony of post-Soviet states is that their determined efforts at creating national histories and identities are resolutely carried on as if a real past can be recovered, as if a continuous, unbroken existence of a coherent nation has come down through time. What is not recognized in the rush to nationhood is just how much work by intellectuals, activists, and state administrators goes into the forging of new nations.

Notes

A longer version of this essay has been published in *The Journal of Modern History*, LXXIII: 4 (December 2001), 862–896.

1. The essay has been published in *The Transcaucasus Today: Prospects for Regional Integration, June 23–25, 1997, Edited Conference Report* (Erevan: American University of Armenia, 1998), 51–57; and as, "Living with the Other: Conflict and Cooperation Among the Transcaucasian Peoples," *AGBU News Magazine*, VII: 3 (September 1997), 27–29.

2. Rogers Brubaker and Frederick Cooper, "Beyond 'identity'," *Theory and Society*, XXIX (2000), 4.

3. See, e.g., Benedict Anderson, "Census, Map, Museum," in *Imagined Communities: Reflections on the Origins and Spread of Nationalism* (London: Verso Press, 1991), 163–185; Bernard S. Cohen, *Colonialism and its Forms of Knowledge: The British in India* (Princeton: Princeton University Press, 1996); and James C. Scott, *Seeing Like a State: How Certain Schemes to Improve the Human Condition Have Failed* (New Haven and London: Yale University Press, 1998).

4. Peter Berger and Thomas Luckmann, *The Social Construction of Reality* (New York: Anchor Books, 1966), 89.

5. Natalie Zemon Davis, "Who Owns History?," in Anne Ollila, ed., *Historical Perspectives on Memory. Studia Historica*, LXI (Helsinki: Suomen Historiallinen Seura, 1999), 21.

6. See, in addition to the list of suggested readings appended to this chapter, Ronald Grigor Suny, "Rethinking Soviet Studies: Bringing the Non-Russians Back In," in Daniel Orlovsky, ed., *Beyond Soviet Studies* (Washington: Woodrow Wilson Center Press, 1995), 105–134; Rogers Brubaker, *Reframing Nationalism: Nationhood and the National Question in the New Europe* (Cambridge: Cambridge University Press, 1996); Geoff Eley and Ronald Grigor Suny, *Becoming National: A Reader* (New York: Oxford University Press, 1996); Theresa Rakowska-Harmstone, "The Dialectics of Nationalism in the USSR," *Problems of Communism*, XXIII: 3 (May–June 1974), 1–22.

7. Robert J. Kaiser, *The Geography of Nationalism in Russia and the USSR* (Princeton: Princeton University Press, 1994), 123.

8. Suny describes *korenizatsiia* as "in effect 'affirmative-action programs'," *The Revenge of the Past: Nationalism, Revolution, and the Collapse of the Soviet Union* (Stanford: Stanford University Press, 1993), 109; and Terry Martin employs the concept as a central metaphor in his *An Affirmative-Action Empire: Nations and Nationalism in the Soviet Union, 1923–1939* (Ithaca: Cornell University Press, 2001).

9. Jeremy Smith, "National Hierarchisation and Soviet Nationality Policy from Lenin to Putin," paper presented at the VI ICCEES World Congress, Tampere, Finland, August 1, 2000; see also his book, *The Bolsheviks and the National Question, 1917–1923* (Basingstoke: Macmillan, 1999).

10. Kaiser, *The Geography of Nationalism in Russia*, 248.

11. Sven Gunnar Simonsen, "Inheriting the Soviet Policy Toolbox: Russia's Dilemma Over Ascriptive Nationality," *Europe-Asia Studies*, LI: 6 (September 1999), 1069–1087; Victor Zaslavsky, *The Neo-Stalinist State: Class, Ethnicity, and Consensus in Soviet Society* (Armonk, N.Y.: M.E. Sharpe, 1982).

12. Slezkine, "The USSR as a Communal Apartment, or, How a Socialist State Promoted Ethnic Particularism," *Slavic Review*, 53: 2 (1994), 450.

13. Ronald Grigor Suny, "Provisional Stabilities: The Politics of Identities in Post-Soviet Eurasia," *International Security*, XXIV: 3 (Winter 1999/2000), 139–178.

14. Ibid., 159–162, 171–173.

15. Bhavna Dave, "Politics of Language Revival: National Identity and State Building in Kazakhstan" (Ph.D. diss., Syracuse University, 1996), 3.

16. Ibid., 227.

17. Martha Brill Olcott, *The Kazakhs* (Stanford: Hoover Institution Press, 1987, 1995), 18; Dave, "Politics of Language Revival," 125.

18. N. Nazarbaev, *Ideological Consolidation of Society as an Essential Prerequisite to Kazakstan's Progress* (Almaty: Daewir, 1994), 40; cited in R. Stuart DeLorme, "Mother Tongue, Mother's Touch: Kazakhstan Government and School Construction of Identity and Language Planning Metaphors" (Ph.D. diss., University of Pennsylvania, 1999), 100.

19. Edward Schatz, "The Politics of Multiple Identities: Lineage and Ethnicity in Kazakhstan," *Europe–Asia Studies*, LII: 3 (2000), 492.

20. Ibid., 496.

21. Ibid.

22. Ibid., 496–498.

23. DeLorme, "Mother Tongue, Mother's Touch," 87–89, 94–95.

24. Ibid., 58, 83–84.

25. Pål Kolstø, "Anticipating Demographic Superiority: Kazakh Thinking on Integration and Nation Building," *Europe-Asia Studies*, L: 1 (1998), 53.

26. Dave, "Politics of Language Revival," 229.

27. From the pamphlet issued by the Kazakhstan Republic President's Office, *Kazakstan Respublikasy aelewmettik-maedenij damuwynyng tuzhyrymdasy/Kontseptsiia sotsiokulturnogo razvitiia respubliki Kazakhstana [Concept of Sociocultural Development in the Republic of Kazakhstan]* (Almaty: Kazakstan, 1993), 8–9; cited in DeLorme, "Mother Tongue, Mother's Touch," 107.

28. Schatz, "The Politics of Multiple Identities," pp. 498–502.

29. Razmik Panossian, "The Evolution of Multilocal National Identity and the Contemporary Politics of Nationalism: Armenia and its Diaspora" (Ph.D. diss., London School of Economics and Political Science, 2000), 37–39. On Shant, see Kevork B. Bardakjian, *A Reference Guide to Modern Armenian Literature, 1500–1920* (Detroit: Wayne State University Press, 2000), 195–197, 484–486.

30. Levon Shant, *Azgutiune himk martkayin enkerutian* [Nationality as the Basis of Human Society] (first published in 1922–1923; reprinted in Erevan, 1999); cited in Panossian, "The Evolution of Multilocal National Identity," 37.

31. E. Hovhannisian, *Azgain kaghakakanutian pilisopaiutiune [The Philosophy of National Politics]* (Beirut: Hamazgaini Vahe Setian Tparan, 1979), 166–167; cited in Panossian, "The Evolution of Multilocal National Identity," 38.

32. H.A. Gevorgian, *Azg, azgain petutiun, azgain mshakuit [Nation, National State, National Culture]* (Simferopol: Amena, 1997); cited in Panossian, "The Evolution of Multilocal National Identity," 38.

33. Mark Saroyan, *Minorities, Mullahs, and Modernity: Reshaping Community in the Former Soviet Union* (Berkeley: International and Area Studies, University of California, Berkeley, 1997), 176–178.

34. Stepan H. Astourian, "In Search of their Forefathers: National Identity and the Historiography and Politics of Armenian and Azerbaijani Ethnogeneses," in Donald V. Schwartz and Razmik Panossian, eds., *Nationalism and History: The Politics of Nation building in Post-Soviet Armenia, Azerbaijan and Georgia* (Toronto: University of Toronto, Centre for Russian and East European Studies, 1994), 41–94; James Russell, "The Formation of the Armenian Nation," in Richard G. Hovannisian, *The Armenian People, From Ancient to Modern Times*, vol. I (New York: St. Martin's., Press 1997), 19–36. On the politics of ethnogenesis, see Victor A. Shnirelman, *Who Gets the Past? Competition for*

Ancestors among Non-Russian Intellectuals in Russia (Washington, D.C.: The Woodrow Wilson Center Press, 1996).

35. Armen Aivazian, *Haiastani patmutian lusabanume amerikian patmagrutian mech*, 8.
36. Panossian reports that "one review in the respected *Patma-banasirakan handes* went so far as saying 'the American authors mentioned in Aivazian's book not only are pro-Turkish in their thinking, *but directly take its false formulations*, explicitly defend them and *act as the lawyers* [of this approach].'" "The Evolution of Multilocal National Identity," 133, n. 317.
37. Ibid., 10.
38. Ibid., 18.
39. Ronald Grigor Suny, *Looking Toward Ararat: Armenia in Modern History* (Bloomington and London: Indiana University Press, 1993), 5.
40. Ibid.
41. See, e.g., Anny Balakian, *Armenian-Americans: From Being to Feeling Armenian* (New Brunswick, NJ, and London: Transaction Publications, 1994); and Susan Paul Patti, *Faith in History: Armenians Rebuilding Community* (Washington and London: Smithsonian Institution Press, 1997).
42. Stephanie Platz, "Pasts and Futures: Space, History, and Armenian Identity, 1988–1994," (Ph.D. diss., University of Chicago, 1996), 17.
43. Ibid., 142.
44. Ibid., 256.
45. Clifford Geertz, ed., *Old Societies and New States: The Quest for Modernity in Asia and Africa* (New York: Free Press of Glencoe, 1963); David E. Apter, *Ghana in Transition* (New York: Atheneum, 1963); James S. Coleman, *Nigeria: Background to Nationalism* (Berkeley: University of California Press, 1958); Reinhard Bendix, *Nation-building and Citizenship: Studies of Our Changing Social Order* (New York: Wiley, 1964).
46. See the now classic article, Dankwart A. Rustow, "Transitions to Democracy: Towards a Dynamic Model," *Comparative Politics*, II (1970), 337–363.

Suggestions for Further Reading

Kaiser, Robert J., *The Geography of Nationalism in Russia and the USSR* (Princeton: Princeton University Press, 1994).

Laitin, David D., *Identity in Formation: The Russian-Speaking Populations in the Near Abroad* (Ithaca, NY: Cornell University Press, 1998).

Martin, Terry, *An Affirmative-Action Empire: Nations and Nationalism in the Soviet Union, 1923–1939* (Ithaca, NY: Cornell University Press, 2001).

Olcott, Martha Brill, *The Kazakhs* (Stanford: Hoover Institution Press, 1987, 1995).

Schwartz, Donald V. and Panossian, Razmik, eds. *Nationalism and History: The Politics of Nation building in Post-Soviet Armenia, Azerbaijan and Georgia* (Toronto: University of Toronto, Centre for Russian and East European Studies, 1994).

Shnirelman, Victor A., *Who Gets the Past? Competition for Ancestors among Non-Russian Intellectuals in Russia* (Washington, D.C.: The Woodrow Wilson Center Press, 1996).

Slezkine, Yuri, *Arctic Mirrors: Russia and the Small Peoples of the North* (Ithaca, NY: Cornell University Press, 1994).

Smith, Jeremy, *The Bolsheviks and the National Question, 1917–1923* (New York: St. Martin's Press, 1999).

Suny, Ronald Grigor, *Looking Toward Ararat: Armenia in Modern History* (Bloomington and London: Indiana University Press, 1993).

Suny, Ronald Grigor, *The Revenge of the Past: Nationalism, Revolution, and the Collapse of the Soviet Union* (Stanford: Stanford University Press, 1993).

CHAPTER 6

Knowledge for Politics: Partisan Histories and Communal Mobilization in India and Pakistan

Subho Basu and Suranjan Das

In late April 2003, a number of historians working on India in the United States received an online petition against the appointment of Romila Thapar, a historian of ancient India, to the Kluge Chair at the Library of Congress. It may seem surprising that such an academic appointment can generate political heat in the expatriate Indian community. Yet the appointment sparked off an orchestrated political campaign on the internet. The petition, entitled "Protest US Supported Marxist Assault Against Hindus," argues that Romila Thapar is responsible for cultural genocide against Hindus.[1]

The question that immediately comes to mind is: why do Hindu nationalists need to devote so much energy to combat the appointment of a historian engaged in research in "classical antiquity"? The answer lies in the pamphlet itself. Hindu nationalists fear that a critical scrutiny of ancient Indian history will disrupt convenient myths about the origins of "Hindu civilization." They regard the earliest episodes of Indian history as sacrosanct and beyond critical empirical investigation, except for eulogistic self-glorification. For them, not only was ancient India a Hindu nation, but it remained unchanged from the time of the composition of the Vedic hymns (nearly one thousand years BC) to the advent of Islam in the Indo-Gangetic plain of South Asia (AD 1206).

Critical examination of ancient Indian civilization poses a threat to the idea of a monolithic "Hindu nation" in India from time immemorial in its unchanging modular form.[2] It was thus feared that the appointment of a nonconformist ancient Indian historian to a visible position within U.S. academia would undermine the cause of Hindu nationalism in the West. The petition cited above, couched in a provocative rhetoric of cultural nationalism and suffused with anti-imperialist references, is a classic example of how the partisan use of the historian's craft can play a significant role in the politics of competing nationalism in South Asia.

As in other multiethnic Asian and African countries that experienced western colonialism, in the Indian subcontinent, too, colonial interpretations of history have informed competing strategies of political mobilization during the struggle for freedom. But there was a qualitative difference in the use of history in mainstream nationalist politics and in sectarian politics. While the former primarily fell back on the "Indian past" to counter the colonial discourse of the superiority of the West over the East, the latter—primarily religious nationalists, both Hindu and Muslim—used history to generate a monolithic identity of religiously based nation-states in a decolonized South Asia. In the postcolonial period, against the background of the political turbulence surrounding diverse trajectories of the two successor nation-states—democratic India, with a Hindu majority but supposedly secular, and Islamic Pakistan—partisan history has continued to play a role in political rhetoric and strategies of mobilization.

This essay explores how Hindu and Muslim nationalists have invoked history in the Indian subcontinent both during and in the aftermath of the freedom struggle. These partisan histories have their origins in the way colonial rulers viewed India, as a complex mosaic of static, unchanging, and conflicting well-defined ethnic communities; the British perceived their role as a neutral umpire controlling these hostile and competing communities. Drawing on colonial historiography that celebrated such essentialized categories of religious, caste, linguistic, and ethnic identities, Hindu and Muslim nationalists developed competing versions of nationalism premised on the belief that Hindus and Muslims constituted two distinct nations that could not coexist.[3] While Hindu nationalists looked to transform Hindu civilization into an exclusivist nation-state, their Muslim counterparts invoked a modular partisan past to justify the separatist movement for a Muslim homeland in the shape of Pakistan, and then to provide an ideological legitimacy for an essentially Islamicized militaristic nation-state. This is how, as the essay hopes to demonstrate, history itself became the site of power struggles based on religious nationalism in South Asia.

Colonial and Nationalist Historiography

In a classic statement of a Eurocentric philosophy of history, the German philosopher Hegel, otherwise an admirer of Indian civilization, asserted that India had no history, in contrast, for example, to China, or of course Europe. The underlying assumption here was that whereas Europe, the birthplace of modernity, has a well-defined historical consciousness, the formerly colonized non-European world actually comprised people without historical consciousness.[4]

Most British historians in colonial India were important bureaucrats who extended their historical assumptions into the strategies of governance, thus carefully transforming their discourses into a new epistemological system of the past to project colonial power and domination. Drawing upon Persian texts, early colonial rulers introduced in the English language survey histories that became a biography of the Indian civilization.[5] In these works, the periodization of Indian history followed the contemporary European ideals of three stages of historical evolution—classical antiquity, dark medieval era, and the post-renaissance modern period. The only difference in the Indian context was that the tripartite epochal division of history became linked with the religion and nationality of ruling elites. Ancient India was referred to as the Hindu period, ignoring the period's complex religious systems. In a similar fashion, medieval India was identified with Muslim rule, although there were powerful non-Muslim rulers in that period as well. The Muslims in India were also treated as a monolithic community, as if all of them were migrants and invaders, claiming descent from the Arabs, Turks, Afghans, and Mughals.[6] The colonial phase of Indian history, in turn, was looked upon in colonial historiography as a harbinger of universal modernity through evangelicalism, western science, philosophy, and rationality.

Colonial historical writings also concentrated on the story of Hindu resistance to "Muslim rule." Historical sources were also compiled in a manner to drive home this point, as is best illustrated in Elliot and Dowson's *History of India As Told By Her Own Historians* (London, 1866–1867). This collection of bardic tales and a romantic reconstruction of Rajput and Maratha rebellions against Mughal rulers presented as authentic academic history captured the imagination of newly rising Hindu middle classes, and became a source of antagonism between Hindus and Muslims.

Another aspect of this discordant Hindu–Muslim relationship in colonial historiography was the semiotics of sexuality. For example, one tale collected by a nineteenth-century British scholar told of the lust of Alauddin, a powerful Muslim monarch, for a beautiful Hindu Rajput queen, and hence his desire to conquer the Hindu Kingdom. This contributed to a stereotype of Muslims as rapists, and Hindus as passive, docile, and submissive.[7] The text romanticized the collective suicide of Hindu women to evade the lustful Muslim men, and thus justified the burning of women on funeral pyres of their husbands. Women were denied individuality and choice, and were posited as the representatives of the community's honor. The image of masculine Muslim and effeminate Hindu became the source of anxiety for Hindu nationalists, who in response preached a militarized spirituality for Hindus.[8] This quest for militarized spirituality gave birth to Hindu nationalist organizations to combat putative "Muslim aggression."

Colonial historiography also racialized Hindu identity. In this, it followed comparative philology, which asserted similarities between Sanskrit and ancient Germanic and Latin linguistic structures, positing a common origin of Aryan civilization in India, Iran, and Western Europe.[9] In a detailed investigation of the origin and impact of this history, Vasant Kaiwar wrote:

> The study of Indo-European languages across the whole space of Europe and south West and south Asia established the idea of diffusion of the master culture and civilization of the Aryans from a central homeland. In this model, Greece and India served as the opposing poles of the dialectic of world history. Both were neatly detached from their local contexts and attached to a central diasporic model, with classical Greece anchoring a triumph list account of European history, and India—after a brief romantic flirtation with the notion of an Oriental renaissance—illustrating a story of decline and degeneration, except in India itself, where the Aryan model and ideas generated by the Oriental renaissance could be mobilized behind conservative agendas of national renaissance.[10]

Partisan History and Hindu Nationalist Discourse

As the rise of mass nationalism against the Raj was equally matched, in the early twentieth century, by an intense Hindu–Muslim communal divide, both Hindu and Muslim nationalists developed competing narratives of the history of their respective communities. The lead in this direction came from the founding ideologue of Hindu nationalism—Vinayak Damodar Savarkar. Nurtured by the tradition of masculine Hindu radicalism and informed by brahmanical social conservatism that rejected the moderate, cosmopolitan, and constitutional politics of the dominant Congress Party, Savarkar became involved with revolutionary politics during his student days in India and England, which led to frequent imprisonment. He wrote a nationalist account of the 1857 Revolt, celebrating it as the First War of Indian Independence. Following the breakdown of Hindu–Muslim solidarity, which had been cemented around the Khilafat–Non Cooperation Movement and the subsequent communal violence, Savarkar emerged as a prominent organizer of the Hindu nationalist party, the Hindu Mahasabha.

Through his writings Savarkar went beyond the traditional invocation of Hindu sensibilities and created the ideological foundation of the Hindu supremacist movement by arguing for the desemitization of Indian culture. He developed the concept of Hindutva (Hinduness), which stipulated that Indians could be those whose *pitri-bhumi* (land of their ancestors) and *punyabhumi* (land of their religion) lay within the territory of British India. Since Muslims and Christians supposedly regarded Arabia and Palestine as their holy land, they could not claim the rights of citizenship.[11] By the same logic the communists, indoctrinated with their atheism and proletarian internationalism, were to be denied Indian citizenship. Such an idea of an exclusivist Indian citizenship was further explicated by another Hindu nationalist theoretician, M.S. Golwarkar, who categorically stated that non-Hindus could never be Indian citizens.[12] Savarkar believed that the idea of Hinduness could not be defined in clear terms, but it could be summed up as the entirety of the history of Indian civilization. In *The Six Glorious Epochs of Indian History*, which many of his followers regard as an elucidation of the survival strategy of the Hindu race and nation, Savarkar even praised the persecution of Buddhists in ancient India:

> Pushyametra and his generals were forced by the exigency of the time, when the war was actually going on, to hang the Indian Buddhists who were guilty of seditious acts, and pulled down monasteries which had become the centres of sedition. It was just punishment for high treason and for joining hands with the enemy, in order that Indian independence and empire might be protected. It was no religious persecution.[13]

He also categorized Ambedkar, the leader of the Dalit (untouchable) movement in India, as a "man burning with hatred against Hinduism." When Ambedkar advised his followers to embrace Buddhism to escape from the Hindu caste system, Savarkar claimed that Buddhism aggravated untouchability in ancient India. He employed the colonial stereotype of aggressive Muslim tyranny in medieval India:

> Intoxicated by this religious ambition, which was many times more diabolic than their political one, these millions of Muslim invaders from all over Asia fell over

India century after century with all the ferocity at their command to destroy Hindu religion which was the life blood of the nation.[14]

Sarvarkar did not merely present Muslim men as lustful, but depicted rape and molestation of Hindu women as deliberate ploys adopted by medieval Muslim rulers to demoralize Hindu resistance. Modhamed Tavakoli has coined the term "hate mysticism" to describe Islamic radicals' anger with the West and modernity.[15] Savarkar's hate mysticism formulated a discourse of Hindu supremacism that obliterated the boundary between evil and good, and saw the world from an instrumentalist, nationalist political understanding.

It is this instrumentalist understanding of the Indian past that prompted Savarkar and his associates to admire and praise German Nazism and Italian fascism. On many occasions, he stated that Hitler was the best for Germany. Savarkar and the Hindu Mahasabha celebrated the revival of Aryan culture in Germany. The Nazi newspaper *Volkischer Beobachter* reported on Savarkar's speeches in exchange for the promotion of Germany's anti-Semitic policies.[16] Despite his admiration of Hitler and Germany, Savarkar supported the idea of the establishment of Israel not only in accordance with his theory of nationalism, but he also saw in the Jewish state a bulwark against the Islamic Arab world.[17] A combination of masculine nationalism and hate mysticism transformed Savarkar's doctrine into an Indian version of aggressive religious nationalist political philosophy.

Much of the Hindu ultranationalist history was political rhetoric that in its attempt to inferiorize others did not bother about evidence. Professional historians of India before and after Independence raised questions about the evidential basis of such historical writings. Yet, in popular imagination legends could easily pass for history; thus, it is often difficult to clearly distinguish not just between history and mythography, but also between history and hate literature.

Muslim Nationalism and the Use of History

The Hindu nationalist discourse naturally provoked a Muslim nationalist response, although the latter was not as systematic and organized as the former. Muslim nationalist discourse essentially stressed the emergence of the Muslim community in India as an exclusivist political entity, quite distinct from the majority Hindu community; the demand for a separate homeland for Muslims in the form of Pakistan followed logically from this discourse. The idea of Pakistan originated in England when Choudhary Rahmat Ali (1895–1951) coined the term Pakistan as a homeland for the 30 million Muslims inhabiting British India in a memorable pamphlet *Now or Never; Are We to Live or Perish Forever?* The homeland movement for Indian Muslims that developed under the stewardship of the Muslim League leader Mohammad Ali Jinnah was largely based upon constitutional bargaining, and informed by a modernist dream for parity of Indian Muslims with the Hindu majority within India. In the tradition of South Asian politics, constitutional categories were transformed into political entities through careful emotional investment. This political mobilizational strategy required a powerful rhetoric of the other, fear psychosis, and romantic historical narrative. In tune with Hindu nationalist rhetoric, this line of thought was crucially linked to the idea of two monolithic and homogeneous religious communities coexisting in a situation of nearly perpetual conflict, and sometimes in uneasy peace. This rhetoric constituted the foundational myth of the Islamic state of Pakistan.

Though British India was predominantly non-Muslim, the largest numbers of Muslims in the world were concentrated in the Indian subcontinent. Muslim South Asians, like their Hindu counterparts, never constituted a monolithic homogeneous community. They were internally divided by language, caste, and even different forms of religious practices. However, in north India, particularly in the Indo-Gangetic heartland, the Muslim service elite of the Mughal Empire shared a common memory of successive Muslim empires. Even in the twilight years of the Mughal Empire, they exercised a stranglehold over government jobs and legal professions. The gradual erosion of the Mughal Empire and the Indian princely states undermined the social and economic status of this service elite.

The uneven nature of the progress of colonial modernity, the emergence of new Hindu service elites and competition for limited resources caused a growing feeling of insecurity for this elite. The rise of enumerated identity through the census and growing assertiveness of Hindu nationalism further escalated the sense of uncertainty that enveloped this community. Many Muslim elites sought to adopt colonial modernity and embraced colonial education; concerned about Hindu majoritarian tendencies within Indian nationalist politics, some sought protection through separate electorates and safeguards for Muslims in access to jobs and education. The traditional Muslim intelligentsia—primarily the learned Ulema—however, found it difficult to accept the so-called march of modernity. They increasingly turned toward pan-Islamism and accommodation with Indian nationalism. These binary categories of modernist and traditional intelligentsia are undoubtedly problematic and do not always neatly dovetail with social formations.

With the rise of mass nationalism after World War I, the pan-Islamists joined hands with Gandhian nationalists. This temporarily displaced constitutionalists like Jinnah. But they came back to the limelight as methods of Gandhian mass politics failed to immediately dismantle the Raj. Jinnah now sought to play upon the fears and aspirations of the minority Muslim elite of the heartland. It is here that the nostalgia for the Islamic past was recreated, and the scepter of Hindu domination was presented in a threatening manner. To rally the much-divided Muslim provincial politicians in the majority provinces into an all–India platform Jinnah placed stress on federalism, and the protection of the interests of Muslim majority provinces. This gave him a powerful position as the "sole spokesman" of the Indian Muslims.[18] Yet the brittleness of Muslim politics constantly plagued Jinnah's project and he thus sought solace in the Muslim League's Pakistan Resolution of 1940. The resolution imagined in distinct terms India's two largest religious communities—Hindus and Muslims—as nations, and demanded that areas where Muslims were numerically in a majority should be grouped to constitute an "Independent State/States" in which the constituent units would be accorded considerable autonomy. The historical negotiations over the transfer of power from British to indigenous hands and the subsequent communal strife were presented by Muslim nationalists as a vindication of the assertion in the 1940 resolution that Hindus and Muslims constituted two monolithic religious nationalities, which could not coexist within one territorial unit. The idea of Pakistan could not thus accommodate any historical reading that could suggest a shared cross-communal identity.

Both Hindu and Muslim nationalist discourses cited above had thus fallen back upon a colonial historiography that rejected multiple identities in subcontinental society based on occupation, language, caste, social hierarchy, and religion. Religious nationalisms premised their master narratives of Indian history on the notion of competing religious identities in Indian society. As we have seen, Hindu nationalist discourse advanced the theory that Muslim invaders destroyed Hindu civilization, created a tyrannical form of government, and caused backwardness in Indian society and economy, opening a path for British colonization. But the Hindutva theory, equating Indian culture with Hindu culture, is ahistorical, since what we know today as the Indian culture is an amalgam of several strands—Hinduism, Islam, Christianity, Zoroastrianism, Persian, Greek, Roman, Parthian, Scythian, and so on. In fact, the theory of Hindutva goes against the eclectic nature of Hinduism itself. In its pristine form, Hinduism is a heterogeneous admixture of diverse practices, rituals, and cults, lacking a rigid structure and organized neither around a single text nor a church. But complexities of historical events were sacrificed on the altar of partisan history for the sake of ideological construction of a homogeneous Hindu nation resisting a monolithic Muslim imperial state. Similarly, the Muslim nationalist dictum ignores unique features of Indian Islam, which has appropriated many rituals of other religions in India, including Hinduism, so much so that one scholar has spoken of Islamic syncretism in the subcontinent.[19] Even today Muslim villagers often worship the same deity with their Hindu neighbors either for a good harvest or to fight epidemics. Numerous instances can be cited of Muslims participating in Hindus festivals and vice versa. Muslim religious figures also contributed much to the development of the Bhakti movement in medieval India that brought together Vedantic and Islamic thinking.

Despite the ahistorical nature of these Hindu and Muslim nationalist discourses, by the 1940s they had brought about a political stalemate in the progress toward independence, as communal fears and apprehensions were channeled against specific individuals, institutions, objects, and symbols. The Pakistan movement and its Hindu nationalist counterpart occasioned in India what Scribner has called, in the context of the German Reformation movement, the process of negative assimilation, wherein the common person became convinced of the rightness of the "evangelical cause," and therefore felt compelled to support the movement.[20] This set the stage for the decolonization process in South Asia, in which the departing colonial rulers divided the subcontinent into two successor states along religious lines before quitting the region.

Partisan History and Political Mobilization in Postcolonial South Asia

The closing years of colonialism in the subcontinent were marked by a series of devastating human tragedies: the painful exodus of Indian refugees from Burma in the wake of Japanese invasion, the ruthless suppression of the 1942 August Revolution, and the man-made Bengal famine of 1943 that took a toll of nearly three million lives. The exit strategy from the Indian empire worked out well for India's former imperial rulers, but not for Indians themselves. A conservative estimate indicates that nearly 14 million people were rendered homeless refugees, nearly 500,000 people perished, and 250,000 women were raped and mutilated in the tragic run-up to the Partition of the subcontinent on August 15, 1947. These statistics, as Gyanendra Pandey has shown, do not capture the nature of human tragedy associated with the Partition saga.[21] The brutalization of human consciousness that occurred during that tumultuous period informed the writings of poets, literary figures, historians, and political activists in subsequent decades. In fact, the Partition became the most common reference point for the history in postcolonial India and Pakistan.

To Hindu nationalists, the Partition validated their historical assumptions. They viewed the event as an evidence of Muslim failure to accept India as their fatherland and blamed the ruling Congress Party and the "effeminate" Gandhian variety of nonviolent nationalism for creating a situation that enabled the British to impose a "truncated settlement." The assassination of Gandhi by a Hindu nationalist was one such expression. Hindu nationalist rhetoric saw Congress leaders as appeasers of the Muslim minority through their ideology of "pseudo-secularism"—evident, for example, in attempts at a democratic solution to the Kashmir problem. Hindu nationalist discourse gradually developed a cultural nationalist project to create a unified corporatist political organizational structure for Hindus, through such bodies as the Bharatiya Jana Sangh (later Bhartiya Janata Party, hereafter BJP), the Rashtriya Swayamsewak Sangh (hereafter RSS) and the Vishwa Hindu Parishad (hereafter VHP).[22]

From the beginning, the writing of Indian history became the ideological battleground between RSS-sponsored organizations and the Indian nation-state under stewardship of the India's first prime minister, Jawaharlal Nehru. While Nehru imagined a new India in terms of social justice, cultural pluralism, and political democracy, Hindu nationalists fell back on Savarkar's dream of a Hinduized state polity. But once the modernist developmental model of the state failed to deliver the promised progress and India fell victim to what has been called "dynastic democracy," the newly enfranchised marginal social groups registered their presence in national politics through regional political forums; in the long run, this helped Hindu nationalist forces to gain a foothold in Indian federal politics.

The BJP-led Hindu nationalist block ensured its political ascendancy by carefully orchestrating the Ramjanmabhumi[23] Movement in the 1980s. This movement aimed to liberate from Muslim stranglehold the birthplace of the Hindu mythical king Lord Rama by removing a mosque called the Babri Masjid that had been built in 1528, allegedly on that sacred spot. The movement—projected as an expression of both Hindu religiosity and Hindu national pride—evoked a passionate explanation of history that combined myths and legends with partisan historical analysis and even archaeological data. In an environment of growing distrust of the ruling Congress Party and the political establishment associated with it, the movement instantly captured the imagination of upper caste and lower middle-class Hindus. The late 1980s was also a period of growing assertiveness of backward castes, which demanded affirmative action to bolster their growing political and economic clout. Alarmed by such developments, India's Hindu upper and lower middle classes extended their support to the Ramjanmabhumi Movement. The result was a tectonic shift in Indian politics.

The Ramjanmabhumi Movement used new technologies such as audio and video cassettes to spread the message of Hindu nationalism. It employed a language of visceral hatred against India's Muslim minorities. Women leaders of the VHP employed the most powerful and populist anti-Muslim rhetoric, grounded in a Hindu interpretation of the Indian past. The most powerful examples of such leaders were Sadhvi Rithambara and Uma Bharati, both of whom were sanyasins—female monks who take the vow of chastity. Hinduism regards women's sexuality as both powerful and dangerous.[24] A celibate woman could sublimate her femininity and be free from the constraints of social restrictions and could thus command the respect of the people for her sexual purity and emancipated social position. These women criticized the Indian political establishment for being "eunuchs," submitting to Muslim pressure. Rithambara constantly referred to Muslim threats to ordinary Hindu women and their sexual purity, and evoked memories of the pre-Partition communal turmoil. In her speeches she referred to the Partition as the vivisection of mother India, describing the present Indian political map as a country without arms. In her speeches, she employed such melodic couplets as:

Muslims, like a pinch of sugar, should sweeten a glass of milk; instead, like lemon, they sour it. What they do not realize is that a squeezed lemon is thrown away while the milk that has been curdled solidifies into paneer [cheese]. So Muslims have two choices, either to live like sugar or like wrung lemons.[25]

Such verses thus implied that the Muslims were to choose between assimilation with Hindus and death. Uma Bharati, the other woman leader of the VHP, was more explicit in referring to Muslims, such as in a 1991 election speech:

Declare without hesitation that this is *Hindu rashtra*, a nation of Hindus. We have come to strengthen the immense Hindu shakti [force] into a fist. Do not display any love for your enemies... The Qur'an teaches them to lie in wait for idol worshippers, to skin them alive, to stuff them in animal skins and torture them until they ask for forgiveness... [We] could not teach them with words, now let us teach them with kicks... Tie up your religiosity and kindness in a bundle and throw it in the Jamuna... A non-Hindu who lives here does so at our mercy.[26]

The Ramjanmabhumi Movement successfully generated a wave of anti-Muslim hatred that culminated in the pulling down of the Babri Masjid on December 6, 1992. The incident provoked an orgy of communal riots across the length and breadth of the country. Members of Muslim and Christian minorities were targeted and murdered systematically through carefully orchestrated mob frenzy. A partisan use of Indian history remained at the center of the Hindu nationalist movement as its leaders expanded their targets to several other "controversial" Muslim religious places supposedly constructed by Mughal emperors to subjugate Hindu aristocracy. Riding on the crest of Hindu nationalism, the BJP-led National Democratic Alliance wrested power in New Delhi, holding power until May 2004.

In a bid to buttress its ideological hegemony, this BJP-led government embarked upon an ambitious project to rewrite history textbooks for school children with a distinct Hindu bias.[27] To prove the genesis of Indian society as Vedic Hindu, even the Indus Valley civilization was claimed "to have been authored by the Vedic Aryans," although this is "untenable on the existing linguistic evidence."[28] Many documentation projects that had been initiated by such institutions as the Indian Council of Historical Research were abruptly terminated on grounds of anti-Hindu and Left bias. This bid to revise history—not unlike the Nazi abuse of archaeology in the 1930s to prove the Aryan origins of the German race[29]—constituted the core of the Hindu nationalist project, in its battle for control over the ideological meaning of the nation-state. In waging this battle, the BJP-led regime not only adopted the Hindutva discourse of Savarkar, but also the modular structure of history formulated by colonial historiography. At a time when the scope of historical enquiry is being expanded to include the study of economics, technology, state formations, environment, gender, and other subaltern identities, the Hindutva revision of history falls back upon dated colonial historiography. Even the naming of new weapons added to the military arsenal; satellites launched with Hindu names reflected the Hinduization efforts of the then BJP-led political establishment. History was sought to be used to imprison Indians with a mind-set of communal hatred and the idea of a monolithic and exclusivist Hindu past.

This recently ousted Hindu nationalist establishment in New Delhi used rituals and symbols to restore Savarkar, who provided the theoretical foundations for contemporary nationalist political mobilization, to the center stage of modern Indian political thought. On February 26, 2003, the president of the Republic of India, A.P.J. Abdul Kalam, unveiled Savarkar's portrait in the Central Hall of the Indian Parliament. Since Savarkar was not simply the ideological founder of Hindu nationalist movement but was believed to have been involved in conspiring the assassination of Gandhi,[30] the entire opposition in parliament boycotted the unveiling. While Hindu nationalists celebrated the ceremony, the event generated diverse forms of protest. No less a person than the former naval chief Admiral Vishnu Bhagwat openly expressed his doubts about the patriotism of the Hindu nationalist leader. In retaliation, the cadres of the Shiv Sena, a local Hindu nationalist organization, physically attacked him. Hindu nationalist organizations even demanded that state governments (the federative units of the Indian union) should install the portrait of Damodar Savarkar in state legislative assemblies.

Even before the unveiling of Savarkar's portrait in the Indian Parliament, the government in New Delhi had performed another symbolic act to rehabilitate the Hindu nationalist leader in the Indian political psyche. On May 5, 2002, the then home minister L.K. Advani renamed the Port Blair airport in the Andaman Islands as Veer (Valiant Hero) Savarkar Airport. Savarkar himself had been a political prisoner in the Andaman prison, a place similar in symbolism to Robben Island Prison in South Africa; Hindu nationalists sought, in this way, to establish their claim to Indian nationalism. The very reference to Savarkar, however, provoked a nationwide criticism from the former Andaman inmates, who questioned the credentials of "Veer" Savarkar as a true soldier of the freedom movement. Evidence was cited of petitions by Savarkar to the British government in 1911 and 1913 (during his prison sentence) seeking pardon from the British government and allegedly pledging loyalty to the imperial ruler. The 96-year-old Bhagat Singh Bilga, a freedom fighter and a former

Andaman inmate, slammed the central government for naming the airport after a person who, he claimed, flinched before the British authority and aptly condemned the episode as a "dishonour" to those freedom fighters who suffered inhuman torture in the Andaman jails but never flinched.[31] The act of renaming the airport, and the hanging of a portrait thus became occasions for political and ideological battles over the history of the nationalist movement in India.

History was also a powerful tool for the managers of the Pakistani state and its military leaders as they constructed Pakistani nationhood after August 1947. Pakistani nationalism had evolved in the Indian subcontinent not primarily as territorial nationalism that could draw upon territorial patriotism suffused with ethnoreligious meanings, as was evident in the case of the Hindu nationalist discourse. This put the founding fathers of Pakistan in a difficult situation. Historically speaking, the center of Indo-Islamic power lay in the north Indian Gangetic plain, in the fertile region between the Yamuna and Ganga rivers, and had never been located in the region that came to constitute Pakistan. The imperial cities of the Indo-Islamic empires such as Delhi, Agra, and Lucknow were located in the heart of India. In fact, to the great irritation of Pakistani leaders, the managers of the Indian nation-state made subtle symbolic claims to the Mughal heritage by holding Independence Day celebrations in front of the Red Fort, a palace constructed by the Mughal emperor Shahjahan between 1618 and 1647. This made it difficult for the Pakistani ruling elites to use the grand Mughal architectural heritage to reclaim symbolically the sub-continent's imperial Islamic past.

Partisan history premised on Muslim nationalist discourse of the days of nationalist struggle became more important in Pakistan because of the absence of political legit-imacy for its military rulers. Intense regional feuds, the absence of an overarching shared notion of history, and confusion over the main ideological content of the nation-state (i.e., Islam versus a secular Muslim identity), also contributed to the importance of history in political conflict. To vindicate the stand that Pakistan was a nation of South Asian Muslims, the founding fathers of the state adopted Urdu as the national language, although no Pakistani community, except for the immigrants from the Indian states of Bihar and Uttar Pradesh, spoke Urdu as their first language. The idea of Bengali as the language of non-Muslims implied a homogeneous Muslim identity.

The colonial notion of racial inferiority of the indigenous non-Aryan population now returned in the language of Pakistan's military dictators. Indeed, General Ayub Khan, the military ruler of Pakistan (1958–1969), wrote that the Muslims:

> probably belong to the very original Indian races . . . They have been in turn ruled by the caste Hindus, Moghuls, Pathans, or the British. In addition, they have been and still are under the considerable Hindu cultural and linguistic influ-ence. As such, they have all the inhibitions of downtrodden races and they have not yet found it possible to adjust psychologically to the requirements of the new born freedom.[32]

The irony is that it was Ayub Khan who trampled the newly formulated constitution of Pakistan in 1958 through a military coup and ushered in an era of rule by handpicked members from the military, who talked about the newfound freedom of the nation.

Yet what is more noticeable here is the echo of colonial historiography and the belief that the supposed pre-Aryan inhabitants of India constituted inferior vanquished nations. So deep was the belief among Pakistani military officers that Bengali Muslims were "converted Hindus" without a clear understanding of the purity of Islamic faith that one of them claimed during the 1970–1971 civil war:

> This is a war between the pure and the impure . . . The people here may have Muslim names and call themselves Muslims. But they are Hindu at heart. We are now sorting them out . . . Those who are left will be real Muslims. We will even teach them Urdu.[33]

This ideological conflation of religion, language, and masculine Muslim purity drove the Pakistani army to commit horrendous atrocities upon the Bengali people. Yet the denial of equal status to Bengali—a language with heavy inputs of Sanskrit words and reflecting a shared Hindu–Muslim culture—spoken by the Muslims in East Pakistan exposed the brittleness of the Pakistani state and directly contributed to the loss of East Pakistan, which became the sovereign state of Bangladesh. The emergence of Bangladesh undermined the legitimacy of the two-nation theory that relied on a nationalism exclusively based on religious identity.

Nevertheless, Pakistani officials continued to approach history as a frozen instrumental discipline. The subjection of history to state regulation has undermined its independence from the struggle for political power. In state-sponsored history textbooks in Pakistan, the study of Indian history becomes worthwhile only after the establishment of "Muslim rule." Muslim invasions from the seventh century onward are thus celebrated as the beginning of the process of the establishment of a Muslim nation in South Asia. The expansion of Muslim empires is portrayed as the triumph of Islam over Hindu India. Certain Muslim monarchs like Akbar, who are celebrated in Indian textbooks for their religious tolerance and syncretic beliefs, are ignored, while monarchs such as Aurangazeb, who are vilified in Indian textbooks for their Islamic bigotry, are praised in Pakistani texts for their piousness. History came to be written and projected primarily from the perspective of a nation-state and its need to produce a religious nationalist identity. In an interview with the *Times of India*, the noted Pakistan historian Mubarak Ali was quoted as saying:

> history writing had suffered in Pakistan because of fanaticism. Anyone writing against the ideology of Pakistan is liable to be jailed for 10 years, according to the Pakistani Ideology Act of 1991. It is also not possible to write against M.A. Jinnah, the founder of Pakistan . . . He said history writing had not developed as a discipline in Pakistan because of such problems. History writing was confined to writing of text-books and articles in newspapers . . . History books in Pakistan traced the two-nation theory to the time of Akbar who was blamed for the downfall of the Mughals. Akbar was not even mentioned in school books. However, Aurangazeb was praised.[34]

The unleashing of a fresh dose of Islamization of Pakistan by General Zia ul-Haq, who ruled Pakistan under martial law from July 5, 1977 till his death in a plane crash on August 17, 1988, intensified the use of the rhetoric of religious nationalism in history writing in Pakistan. The Afghan crisis and the radicalization of Islam had a further impact on the domestic politics of Pakistan. Today, in the post–September 11 world, Pakistan is still in the eye of the Islamic fundamentalist storm. It is thus no surprise that in 1998 the upgraded Pakistani ballistic missiles—which, with a range of 2,700–3,000 kilometers, can target virtually all important Indian establishments—were named Ghauri, after the twelfth century general Muhammad Ghori, who conquered Delhi in 1206. Partisan history could be powerful enough to convince today's generals to repeat history with upgraded nuclear weapon systems. Yet Pakistan exists today as "Nationalism without a Nation," failing to resolve the competing ethnic claims of Punjabis, Sindhis, Mohajirs, Baluchis, and Pakhtoons.[35]

Conclusion

One can raise the question as to whether a historical piece can be anything but partisan, since we can no longer speak of a purely objective and value-free historical discourse.[36] It has also been argued that all history is contemporary history, given the fact that the understanding of a historian is shaped by the age in which he lives. Yet, following E.J. Hobsbawm, we can make a sharp distinction between subjective partisanship in historical literature and partisan history per se. Whereas the former "rests on disagreement not about verified facts, but about their selection and combination, and about what may be inferred from them," the latter is essentially a recording of the past to serve particular political interests based neither on established methodological canons nor reliable evidence.[37] The colonial, Hindu, and Muslim nationalist discourses considered above fall in the second category. Such histories provide legitimacy to sectarian political action and provide fuel to the formation of contested national identities.

The identity of victim and the desire for revenge so informs partisan history that, with state sponsorship, it can become a real disruptive force. In the hands of its partisan practitioners, the study of history is no longer concerned with an exploration of knowledge. Rewriting history is essential to keep in touch with the advancing frontiers of knowledge, but rewriting for partisan purposes as experienced in India and Pakistan for imprinting the ideologies of ruling regimes, the notion of Hindu rashtra in the case of India and Islamic fundamentalism in the case of Pakistan, transforms history into an instrumentalist discipline. It is also important to recognize, as indicated above, that such historical writings were products of colonial modernization. This very colonial project has survived in the form of religious nationalist ideologies trying to foster essentialized religio-ethnic identities both before and after the colonial period. The modern globalized structure of politics plays a crucial role in sustaining and disseminating such religious nationalism. Thus, it is not without significance that the rising strength of religious nationalism in both India and Pakistan has been contemporaneous with the blatant subjection of the two states to the force of globalization. This is perhaps because, as Romila Thapar aptly remarks: "The new communities created by globalization are supposed to be modern but where modernization fails them they use religious identities as a cover for a barbaric cult of terror and fear."[38] Both in India and Pakistan, this process has been sustained by the financial support of their residents settled abroad.

The victims in each case have been democracy, secularism, social equity, and justice. Yet democracy can provide a cure for the extremes of religious nationalism. This became evident when in India the ruling BJP-led Hindu fundamentalist coalition that had engaged itself in popular mobilization through its modular partisan history was defeated in the fourteenth parliamentary election of May 2004. The new Indian National Congress–led United Progressive Alliance government has positioned itself in favor of secular democratic politics.[39] In the recent Indian parliamentary election, the electorate, contradicting all predictions of the victory of Hindu nationalists and their ability to mobilize people through their modular partisan history, passed a verdict in favor of an alliance of secular and democratic forces. It reveals that in a democratic society, despite the ready-made attraction, partisan history could become a trap for its proponents as well.

Notes

We would like to record our thanks to Prof. Bhaskar Chakrabarty of Calcutta University for suggesting crucial editorial changes in the initial draft of this chapter.

1. Available at http://www.petitiononline.com/108india/petition.html. The petition reads, in part: "she is an avowed antagonist of India's Hindu civilization. As a well-known Marxist, she represents a completely Euro-centric world view. I fail to see how she can be the correct choice to represent India's ancient history and civilization. She completely disavows that India ever had a history . . . The ongoing campaign by Romila Thapar and others to discredit Hindu civilization is a war of cultural genocide."

2. Benedict Anderson uses the notion of "modular"—i.e., capable of being transplanted in diverse social terrains with varying degrees of self-consciousness—in his discussion of print capitalism and the formation of "imagined communities." Anderson argues that nation-ness, as well as nationalism, are cultural artifacts of a particular kind. See Anderson, *Imagined Communities: Reflections on the Origin and Spread of Nationalism* (London: Verso, 1993), 4. Partha Chatterjee questions this "euro-centric" idea, arguing that the notion of a modular imagined community reduces Asians and Africans to perpetual passive consumers of a modernity emanating from European and American sources. Chatterjee, *The Nation and Its Fragments: Colonial and Postcolonial Histories* (Princeton: Princeton University Press, 1993), 4–5.

3. See Gyan Prakash, "Writing Post-Orientalist Histories of the Third World: Perspectives from Indian Historiography," in Vinayak Chaturvedi, ed., *Mapping Subaltern Studies and the Post Colonial* (London and New York: Verso, 2000).

4. For detailed discussion of the theme, see Eric Wolf, *Europe and the Peoples without History* (Berkeley and Los Angeles: University of California Press, 1997).

5. Examples are James Mill, *The History of British India* (4th edn. with notes and contribution by H.H. Wilson London: 1840–1848); Sir William Jones, *Discourses delivered before the Asiatic Society and miscellaneous papers on the nations of India with an essay by Lord Teignmouth* selected and edited by J. Elmes (London: 1824, 5 vols.); and Vincent Arthur Smith, *The Oxford History of India* (Delhi: 1981 ed.). On this historiography, see Romila Thapar's submission on "History as Politics," available at www.indiatogether.org/2003/may/opl-history.htm.

6. Recent research, in fact, has shown that a substantial majority of Indian Muslims were converts from Hinduism. It has also been argued that contrary to colonial projection of the clash between Muslim invaders and indigenous power blocks in terms of religious conflict, the issue at stake was primarily territory, political power, and status. See Thapar, "History as Politics."

7. Discussed in Purushottam Agarwal, "Savarkar, Surat and Draupadi: Legitimising Rape as a Political Weapon," in Tanika Sarkar and Urbashi Butali, eds., *Women and the Hindu Right: A Collection of Essays* (New Delhi: Kali for Women, 1996).

8. See Indira Chowdhury, *The Frail Hero and Virile History: Gender and the Politics of Culture in Colonial Bengal* (New Delhi: Oxford University Press, 1996); and Sikata Banerjee, *Warriors in Politics: Hindu Nationalism, Violence, and the Shiv Sena in India* (Boulder: Westview Press, 1999).

9. See Thomas Trautmann, *Aryans and British India* (Berkeley and Los Angeles: University of California Press, 1997).

10. Vasant Kaiwar, "The Aryan Model of History and the Oriental Renaissance: The Politics of Identity in an Age of Revolutions, Colonialism and Nationalism," in Vasant Kaiwar and Sucheta Mazumdar, eds., *Antinomies of Modernity: Essays on Race, Orient, Nation* (Durham and London: Duke University Press, 2003).

11. V.D. Savarkar, *Hindutva: Who is a Hindu?* (Delhi: Bharati Sahitya Sadan, 1989).

12. M.S. Golwarkar, *Bunch of Thoughts* (Bangalore: Sahitya Sindhu Prakashna, 1996).

13. V.D. Savarkar, *The Six Glorious Epochs of Indian History* (Delhi: Rajdhani Granthagar, 1984), 85. Quoted in Agarwal, "Savarkar Surat and Draupadi," 44.

14. Savarkar, *Six Glorious Epochs*, 129–130, quoted in Agarwal, 48.

15. Modhamed Tavakoli, *Refashioning Iran: Orientalism, Occidentalism and Historiography* (Oxford: Palgrave, 2001).

16. Christophe Jaffrelot, *The Hindu Nationalist Movement and Indian Politics, 1925 to the 1990s* (London: South Asia Press, 1996).

17. For a detailed explanation of Savarkar's appreciation of Nazi Germany and fascist Italy, see Marzia Casolari, "Hindutva's Foreign Tie-Up in the 1930s: Archival Evidence," *Economic and Political Weekly*, January 22, 2000. Also see Chetan Bhatt, *Hindu Nationalism Origins, Ideologies and Modern Myths* (Oxford: Berg, 2001), 105–108.

18. Ayesha Jalal, *The Sole Spokesman: Jinnah, the Muslim League and the Demand for Pakistan* (Cambridge: Cambridge University Press, 1994).

19. Asim Roy, *The Islamic Syncretistic Tradition in Bengal* (Princeton: Princeton University Press, 1983); also see his *Islam in South Asia: A Regional Perspective* (New Delhi: South Asia Publishers, 1996).

20. See Suranjan Das, *Communal Riots in Bengal 1905–1947* (Delhi: Oxford University Press, 1990).

21. For an excellent analysis of the violence that accompanied the Partition of British India in 1947, and the memory and "shifting meanings and contours" of that violence, see Gyanendra Pandey, *Remembering Partition: Violence, Nationalism And History In India* (Cambridge: Cambridge University Press, 2001).

22. The RSS, founded in 1925, is an organization aiming to establish a Hindu value system in India. Dr. K.B. Hedgewar is considered its founding father. The Viswa Hindu Parishad was founded in 1964 with the following objectives: to establish connections with Hindus settled outside India, to reintegrate within Hindu society all those who had once cut themselves off from it, and to spread Hindu values. Both these organizations are part of the broader Hindu fundamentalist front in India, known in political parlance as the "Sangh parivar." See Thomas B. Hansen, *The Saffron Wave: Democracy and Hindu Nationalism in Modern India* (Princeton: Princeton University Press, 1999), ch. 3.

23. For an overview of the movement see Sarvepalli Gopal, ed., *Anatomy of a Confrontation: The Babri Masjid–Ramjanmabhumi Issue* (New Delhi: Viking, 1991).

24. Amrita Basu, "Feminism Inverted: The Gendered Imagery and Real Women of Hindu nationalism," in Sarkar and Butali, eds., *Women and the Hindu Right*, 162–163.

25. Ibid., 163.

26. Quoted in Hansen, *The Saffron Wave*, 180.
27. See Krishna Kumar "Continued Text"; and Rashmi Paliwal and C.M. Subramanium "Ideology and Pedagogy," in *Seminar*, 400 (December 1992), 17–19, 34.
28. Thapar, "History as Politics."
29. Ibid.
30. See A.G. Noorani, *Savarkar and Hindutva: The Godse Connection* (New Delhi: LeftWord Books, 2002). See also the review of the book by P.K. Datta, in *Frontline* 20: 8 (April 12–25, 2003).
31. See *Deccan Herald*, November 24, 2002.
32. Quoted in Philip Oldenburg, "A Place Insufficiently Imagined: Language, Belief and the Pakistan Crisis of 1971," *Journal of Asian Studies*, 44: 4 (1985), 724.
33. Anthony Mascarenhas, "Genocide," *Sunday Times of London*, June 13, 1971.
34. Interview by Vidyadhar Date, *The Times of India*, January 29, 1999.
35. Christophe Jaffrelot, ed., *Pakistan: Nationalism without a Nation?* (New Delhi and London: Manohar and Zed Books, 2002).
36. This is well elucidated in E.H. Carr, *What is History?* (Penguin: Harmondsworth, 1964 ed.).
37. Eric Hobsbawm, *On History* (London: Abacus, 1990).
38. Thapar, "History as Politics."
39. The present Indian government under the premiership of Manmohan Singh is supported from outside by the 61-strong Left group in the 543-member lower house of the Indian Parliament (Lok Sabha).

Suggestions for Further Reading

Banerjee, Sikata, *Warriors in Politics: Hindu Nationalism, Violence, and the Shiv Sena in India* (Boulder: Westview Press, 1999).

Chatterjee, Partha, *The Nation and Its Fragments: Colonial and Postcolonial Histories* (Princeton: Princeton University Press, 1993).

Hansen, Thomas B., *The Saffron Wave: Democracy and Hindu Nationalism in Modern India* (Princeton: Princeton University Press, 1999).

Jaffrelot, Christophe, *The Hindu Nationalist Movement and Indian Politics, 1925 to the 1990s* (London: South Asia Press, 1996).

Jaffrelot, Christophe, ed., *Pakistan: Nationalism without a Nation?* (New Delhi and London: Manohar and Zed Books, 2002).

Pandey, Gyanendra, *Remembering Partition: Violence, Nationalism and History in India* (Cambridge: Cambridge University Press, 2001).

Roy, Asim, *Islam in South Asia: A Regional Perspective* (New Delhi: South Asia Publishers, 1996).

Sarkar, Tanika and Butali, Urbashi, eds., *Women and the Hindu Right: A Collection of Essays* (New Delhi: Kali for Women, 1996).

CHAPTER 7

Historiophobia or the Enslavement of History: The Role of the 1948 Ethnic Cleansing in the Contemporary Israeli–Palestinian Peace Process

Ilan Pappe

There is a natural inclination among those political scientists and politicians involved in peacemaking to look at the past and memory as an obstacle to progress. They recommend liberating oneself from the past as a prerequisite for peace. This view is entrenched in a wider context of reconciliation and mediation policies that emerged in the United States after World War II. This school of thought was based on a businesslike approach that treats the past as an irrelevant feature in the making of peace.[1] There is a looser but no less important link with modernization theories, which regard tradition and the past as obstacles to a better future; from this perspective, a party that is involved in a conflict and is insistent on rectification of past evils can easily be depicted as representing not only intransigence but also backwardness. This has been, in fact, the fate of the Palestinian voice in the history of peacemaking in the Middle East.

The "progressive" way toward peace, then, is elimination of the past. The businesslike peacemakers only consider a contemporary situation—with its balance of power and realities on the ground—as a starting point for a reconciliation process. This perception also affects the process of "learning" so crucial in the study of peacemaking. Even when such a peace effort results in blatant failure, the renewed effort restarts from a similar point of view: namely, one that neglects to take into account the lessons of the previous phases' failure. Noam Chomsky, noting such a tendency in the Middle East peace process, concluded that the result was a never-ending "peace process" that was not meant to bring peace, but rather provided jobs and preoccupations for a large group of people belonging to the peace industry.[2]

This philosophy has informed the peace process in Palestine ever since 1948, and in particular after 1967. It has destroyed the chances of peace in Israel and Palestine; only the reintroduction of the historical dimension can save the peace effort. The starting point that has been totally neglected, for reasons that are explained later in the chapter, is the year 1948. The events of this year have not only been excluded from the peacemaking process, but their role and significance in the making of the conflict have been ignored. This chapter therefore begins with a short survey of the 1948 ethnic cleansing in Palestine. The second part of this chapter presents the formulation of four guidelines that have underlined the peace process since 1967 and their ahistorical orientation; the last part suggests possible ways of reintroducing the past into the reconciliation effort.

The 1948 Ethnic Cleansing: The Crucial Starting Point

In February 1947, the British Empire ceased to rule Palestine (known as the Mandate) and referred that territory's future to the United Nations. For 30 years, the British government tried to reconcile an impossibly contradictory pledge it had made to both the Zionist movement and the Palestinians. In the 1917 Balfour Declaration, the British government promised to reconcile the natural rights of the indigenous population of Palestine with its pledge to turn the country into a Jewish homeland. This proved an impossible mission. After endless efforts and several cycles of bloodshed, the decision came to leave the future of that troubled country in someone else's hands.

The international body was a very young and inexperienced outfit in those days, and those it delegated to find a solution to the conflict were at a total loss where to begin and how to proceed. A more experienced team was ruled out because cold war politics guided the superpowers of the day to appoint representatives from member states outside the Eastern and Western blocs, and hence with no prior involvement in or knowledge of the question of Palestine. Not surprisingly the deliberations of these member states in the framework of an ad hoc inquiry committee (UNSCOP) reached a deadlock very early on.[3]

The Jewish Agency, the unofficial government of the Jewish community in Palestine, gladly filled the vacuum and exploited Palestinian disarray and passivity at this crucial historical juncture to the fullest. In May 1947, the Jewish Agency handed a map and a plan to UNSCOP. The opening positions of the two sides were diametrically opposed: each wished to control the whole of the country in the post-Mandatory era, and promised to protect the rights of the other group. Palestinian representatives spoke in the name of the native population, and regarded the Zionist movement as a colonialist one; nonetheless, they conceded that most of its members could remain. The Zionist movement, in turn, depicted itself as a national movement redeeming an ancient homeland taken by strangers who nonetheless would be allowed to remain on it in a separate enclave. They thus envisioned the partition of Palestine in a way that would safeguard Jewish survival and protect Zionist plans.

The Palestinian community rebelled against the British for three years, 1936–1939, and in the struggle saw its leadership exiled and fragmented. The leaders remaining inside Palestine were not of the same caliber, but were confident enough of providing their community with a vision for the future. These politicians regarded Palestine as they regarded Egypt or Iraq: an Arab country that would eventually be transferred to its people. As so many other Arab countries were already granted independence, there seemed to be no need for apprehension. This perception, among other reasons, planted indifference and ensured lack of preparation in the face of the diplomatic struggle in the United Nations. The Jewish Agency, on the other hand, had an assertive leadership that had taken all the necessary steps—recommended in the book of international relations in those days—to ensure United Nations and global support for a peace plan tailored to the needs of the Zionist newcomers, one which disregarded the ambitions of the indigenous population.[4]

The May 1947 plan of the Jewish Agency suggested the creation of a Jewish state that comprised 80 percent of Palestine—more or less the Israel of today without the occupied territories. That same November, UNSCOP reduced the Jewish state to 55 percent of Palestine and formulated the plan as UN General Assembly Resolution 181. The Palestinian rejection of the plan—which did not surprise anyone as Palestinian leaders had been opposed to partition ever since 1918—and the Zionist endorsement of it were in the eyes of the international policeman a solid enough base for peace in the Holy Land.

Imposing the will of one side on the other was hardly a productive move toward reconciliation; indeed, rather than bringing peace and quiet to the torn land, the resolution triggered violence on a scale unprecedented in the history of modern Palestine. The U.S. Department of State reached this conclusion in April 1948, after months of bloodshed triggered by the resolution, and suggested another five years for deliberations, but the Jewish lobby in the United States succeeded in convincing President Truman not to withdraw his support for a plan that served the Zionist movement's aspirations in Palestine so well.[5]

In the wake of the general Arab and Palestinian rejection of the UN Peace Plan, the Jewish leadership felt free to return to its May 1947 map; if the Palestinians refused to go along with the Zionist idea of partition, it was time for unilateral action. The map showed clearly which parts of Palestine were coveted for the future Jewish state. The problem was that within the desired 80 percent the Jews were a 40 percent minority (660,000 Jews and one million Palestinians). But this hurdle could also be cleared. The leaders of the Yishuv (the Jewish community in Mandatory Palestine) had been prepared ever since the beginning of the Zionist project in Palestine for such an eventuality. They advocated, in such a case, the enforced transfer of the indigenous population so that a pure Jewish state could be established.

Transfer and ethnic cleansing as a means of Judaizing Palestine had been closely associated in Zionist thought and practice with "historical opportunities": appropriate circumstances such as an indifferent world, or "revolutionary conditions" such as war. This link between purpose and timing had been elucidated very clearly in a letter David Ben-Gurion had sent to his son Amos in 1937. This notion reappeared ever after in Ben-Gurion's addresses to his MAPAI party (the acronym for the party of Eretz Israel Workers, the main labor party of the Jewish community in Palestine) members throughout the Mandatory period, up to the moment when such an opportune moment arose—in 1948.[6]

The Actual Ethnic Cleansing

A kind of civil war began a day after the UN resolution suggested dividing Palestine and authorizing the creation of a Jewish state. In such cases, it is difficult and quite unimportant to establish who fired first; what is clear is that the UN peace plan triggered a bloodbath that overshadowed all the previous violent events in that land. The Palestinian responses, angry and sporadic, exposed their lack of leadership, manpower, weapons, or a plan. The Jewish side was better equipped, far superior in number of fighting men and purposeful.

From December 21, 1947 until March 10, 1948, Jewish forces attacked a limited number of Palestinian villages and Bedouin settlements in a first "cleansing" operation in areas vital for Jewish communication and administration, such as the coastal plain north of Tel-Aviv. Around the mixed Arab–Jewish Palestinian towns, but not in them, major villages were taken, and in some cases massacres took place, for example at Balad Al-Shaykh, near Haifa, where more than 60 civilians were murdered in the beginning of January 1948.

But these were still limited operations that sowed terror among the more well-off Palestinians (roughly 70,000 people), who fled the country in January and February 1948. Another 100,000 villagers and city dwellers, expelled by the Jewish forces, became refugees. And yet this did not create a "valid" Jewish state as predicted by Ben-Gurion.

The sense, therefore, among the Jewish leaders was that something more systematic was called for. On March 10, 1948, the Hagana, the main Jewish underground in Palestine, issued a military blueprint for the expected British evacuation from Palestine, scheduled for May that year. The plan dealt with two issues: first, how to protect the 80 percent of Palestine designated unilaterally by the Jewish Agency as the future state from the threats of attack from neighboring Arab states; and second, what to do with the million Palestinians living within this territory.

The Plan—known as "Plan D," as it replaced three earlier blueprints for the creation of a Jewish state—evaluated correctly, so it seems in hindsight, the inability of the Arab states to fulfill their pledge to save Palestine, although it envisaged tough fighting on several fronts, especially where isolated Jewish settlements existed on the frontiers. Indeed, these are the parts of Palestine where an actual war took place. Elsewhere, there was no real military force that could oppose the Jewish troops. Meanwhile, Plan D ordered the army to cleanse Palestinian areas falling under its control. The Hagana had several brigades at its disposal; each received a list of villages to occupy. Most of the villages were destined to be destroyed; only in very exceptional cases was the army ordered to leave them intact.[7] When taken together, the demolition and expulsion campaign left behind more than 400 villages and 11 towns in ruins, thousands of massacred civilians and a catastrophe the memory of which would feed the Palestinians' national movement ever since—and the denial of which would become the main cause for the ongoing conflict in Palestine.

First Attempts at Peace

In the first two years after the *Nakbah* (the Arab term for "catastrophe" or "disaster," as the events of 1948 are known), there was enough energy left in the United Nations to produce a diplomatic effort to bring peace to the country. This effort culminated in the convening of a peace conference in Lausanne, Switzerland in Spring 1949. This conference was based on UN Resolution 194, a resolution that, like its predecessor, did not refer to the past as a feature in the making of peace. The events of the present were dramatic enough to draw attention to the refugee community as the major issue at hand. In the eyes of the Palestine Conciliation Commission, the UN mediation body that drafted Resolution 194, the unconditional return of the refugees was the basis for peace in Palestine. This was one of three major features of the proposed solution for post-Mandatory Palestine; the other two were a more or less equal division of the country and the internationalization of Jerusalem.

The peace proposal avoided strictly historical questions such as the Jewish past in the region and Jewish claims to Palestine, the colonialist nature of Zionism, or the loss of Palestine as a homeland and civilization. But at least it dealt directly with the human tragedy unfolding in Palestine. This was the last time that international efforts focused on this issue. Loyal to a concept that disregarded the past and its evils, every peace effort since has been based on questions of balance of power and the more hidden interests and agenda of the peacemakers (in most cases American diplomats).

The comprehensive approach of Resolution 194 was accepted by everyone: the United States, the United Nations, the Arab world, the Palestinians, and by the Israeli foreign minister, Moshe Sharett. But the Israeli prime minister, David Ben-Gurion, and King Abdullah of Jordan, wishing to partition Palestine between their two respective countries, won the day in 1949. The distractions of an election year in America, and the cold war in Europe, allowed these two leaders to carry the day and bury the chances for peace as well as nip in the bud one of the few attempts at a comprehensive approach for peace in Israel and Palestine.[8]

Thus, post-Mandatory Palestine was divided between the Egyptians (in the Gaza Strip), the Jordanians (in the West Bank), and the Israelis who received the lion's share (78 percent of historical Palestine). Much of the country was destroyed and most of its indigenous people expelled. The expulsion and destruction kindled a conflict that has burned ever since. The PLO (Palestinian Liberation Organization) emerged in the late 1950s as an embodiment of the Palestinian struggle for return, reconstruction, and restitution. But it was not a particularly successful struggle. The refugees were totally ignored by the international community and the regional Arab powers. Only Egypt's leader Gamal abd al-Nasser seemed to adopt the Palestinian refugees' cause, forcing the Arab League to show at least concern for their plight. As the Arab states' ill-fated maneuvers in the June 1967 war showed, this was not enough or efficient. The Arab leaders adopted a brinksmanship policy, which enabled Israel to attack and annex the rest of Palestine. This policy neither saved the Palestinians nor did it improve the Arab world's relationship with the West. Against this background the peace process in Palestine began.

Toward a Pax Americana

The June 1967 war ended with total Israeli control over ex-Mandatory Palestine. The peace process began immediately after the short war ended and was more overt and intensive than the one following the 1948 war. The early initiators of the process came from the British, French, and Soviet delegations in the United Nations, but momentum soon enough passed to the Americans, who took the initiative as part of their comprehensive and successful plan to exclude the Russians from Middle Eastern agendas.

The basic assumption underlying the American effort was an absolute reliance on the balance of power as the principal prism through which possible solutions should be examined. Within this balance of power, Israel's superiority was unquestioned after the war; hence whatever came from this side in the form of a peace proposal would serve as a basis for a Pax Americana. Thus, the Israeli peace camp—that is, those in Israeli politics who presented ostensibly the more moderate position toward prospective peace in Palestine—would articulate the conventional wisdom of the peace process and formulate its guidelines.

These guidelines, born in response to the new geopolitical reality that emerged after the June war, were subsequently drafted in a clearer form. This new reality paralleled the debate within Israel between the right wing, which sought to create a Greater Israel including the occupied territories, and the left wing, the Peace Now movement. The former were "redeemers"—people who regarded the areas occupied by Israel in 1967 as the regained heartland of the Jewish homeland; the latter were "custodians," who believed the territories could be used as bargaining chips in future peace negotiations. The "redeemers" controlled Israel in 2005, more than 35 years after the occupation, but the "custodians" continue to dominate the peace agenda.

After these guidelines were adopted by the American apparatus responsible for shaping U.S. policy in Palestine, they were described with such positive phrases as "concessions," "reasonable moves," and "flexible positions." But these guidelines catered to the internal Israeli scene and as such totally disregarded Palestinian points of view—of whatever nature and inclination.

The Exclusion of the Past in the Name of Peace

The first guideline that emerged from the American-backed Israeli position was that the Israeli–Palestinian conflict began in 1967; hence, the essence of its proposed solution is an agreement that would determine the future status of the West Bank and the Gaza Strip. In other words, the 78 percent of Palestine included within the pre-1967 Israeli borders was not open to negotiation, and a solution should be confined to the remaining 22 percent, over which a compromise should be found accordingly in a businesslike manner. The second guideline is that everything in the disputed areas—the land, its natural resources, and the population—is divisible and that such divisibility is the key for peace. The third guideline is that anything that happened until 1967, including the Nakbah and the ethnic cleansing, is not at all negotiable. The implications of this guideline are clear. It removes the refugee issue from the peace agenda completely, and treats the Palestinian right of return as a nonstarter. The last guideline makes an equation between the end of Israeli occupation and the end of the conflict. This is a natural conclusion from the previous guideline.

These guidelines became viable once a potential partner for such a peace was found. At first, these principles were offered to King Hussein of Jordan, with the help of the mediation skills of the American secretary of state, Henry Kissinger. The Israeli peace camp, led by the Labor Party, regarded the Palestinians as nonexistent. This attitude was manifested in the (in)famous declaration by the Israeli prime minister that "there is no such a thing as a Palestinian people." According to this view the Arabs of Palestine were either Jordanians, Syrians, or Lebanese.[9] Labor preferred to divide the occupied territories with the Jordanians. But the share offered to the king in 1972 by Prime Minister Golda Meir, who coveted the whole area including East Jerusalem, was not enough.

This offer, known later as the Jordanian option, was endorsed by the Americans until 1987, the year of the first *Intifada*, the uprising in the occupied territories. Whereas the failure of the Jordanian avenue in the first years was due to lack of Israeli generosity, in later years a greater obstacle was King Hussein's ambivalence and inability to negotiate on behalf of the Palestinians; meanwhile the PLO enjoyed pan-Arab and global legitimacy (in 1975, the PLO had more legations abroad than Israel had embassies).[10]

A parallel path was offered by Egyptian President Anwar Sadat in his 1977 peace initiative and by Prime Minister Menachem Begin of the Likud Party (in power between 1977 and 1982). They suggested allowing Israeli control over the occupied areas while granting internal autonomy for the Palestinians. It was in essence another version of a partition that would leave Israel in direct control over 80 percent of Palestine, with indirect control over the rest.

The first Intifada squashed the idea of Palestinian autonomy, as it led to Jordan's removal as a partner in future negotiations. These developments convinced the Israeli peace camp to accept the Palestinians as a partner in a future settlement. At first Israel tried, with the help of the Americans, to negotiate peace with the leadership of the occupied territories, which attended the 1991 Madrid peace conference as an official peace delegation. The conference was a reward given by the American administration to the Arab states for their backing of the American military operation against Iraq, and it led nowhere.

The guidelines for peace were rearticulated during the days of the Oslo Accord that began in 1993. There was, however, a novel component in Israeli strategy: the Israelis were now looking for Palestinian partners in the search for their kind of peace in Palestine. And they aimed at the top—to the PLO leadership, at that time sitting in Tunis. The latter were lured into the process by an Israeli promise, incurred in article 5 of the Oslo Accord, that after five years of catering to Israeli security needs—that is, a period during which Israeli citizens would not be targeted by Palestinian guerrilla actions—the main Palestinian demands would be put on the negotiating table in preparation for a final agreement.[11] In the meantime, the Palestinians would be allowed to play at independence. They were offered a Palestinian Authority (PA), decorated with the insignia of sovereignty, that could remain intact as long as it clamped down on any resistance movement against the Israelis. For that purpose, the PA employed five secret services, whose conduct added to the occupier's abuse of human and civil rights, those of the indigenous administration. These services were less busy in protecting the new authority; some were there more to silence criticism.

It took seven years for all concerned to realize that since 1967 two new players had emerged: Jewish and Islamic fundamentalist movements. These two reduced the chances of getting an agreement under the Oslo premises.

Jewish fundamentalist movements had already begun to make their mark in 1967; in subsequent years, they gradually dotted the occupied territories with Jewish settlements, located in such a way as to prevent any Palestinian geographical continuity or access to water resources. Islamic fundamentalism, in turn, radicalized Palestinian public opinion and introduced terrorist tactics against Israel, which reciprocated with a brutal repertoire of collective punishments. Meanwhile, apart from the quasi-autonomy, the occupation continued with force—sometimes indirectly, as in the famous areas A (those areas, which according to the Oslo agreement were under ostensibly direct Palestinian control), and in other places directly. The daily harassment of occupation was felt everywhere, as more Jewish settlers flowed into the areas. When the Palestinian opposition retaliated with the deadly weapon of suicide bombs, the Israelis enhanced their repertoire of collective punishment—house demolitions, assassinations, closures, and expulsions—in a way that weekly doubled the support for suicide bombers and their senders. This cycle of violence disabled the emergence of any serious peace camps on both sides of the divide.[12]

Six years after the signing of Oslo, the stormy Israeli political scene brought to power once more the Peace Camp, with Ehud Barak at its head. Just one year later, he was facing electoral defeat for being overambitious in almost every field of governmental policy. Peace with the Palestinians seemed to be the only salvation.

The Palestinians expected the promise made in Oslo to be the basis for new negotiations. As they saw it, because they had agreed to wait so long, now was the time to discuss the problem of Jerusalem, the fate of the refugees, and the future of the settlements. The Israelis, drawing upon academics and "professional" experts to a greater extent than before, once more provided a plan. The fragmented Palestinian leadership was unable to do the job itself and sought advice for alternate plans in such unlikely places as the Adam Smith Institute in London. No wonder that it was the Israeli plan, endorsed by the Americans, that was exclusively on the negotiating table at Camp David in the summer of 2000. The plan offered withdrawal from most of the West Bank and the Gaza Strip, leaving as Palestinian about 15 percent of the original Palestine in the form of detached and distanced cantons bisected by highways, settlements, army camps, and walls. The plan did not envision the capital in Jerusalem, offered no solution to the refugee problem, and represented total abuse of the concept of statehood and independence. Even the fragile Arafat, who had seemed hitherto to be happy with the *salata* (the perks of power) and who did not exercise *sulta* (actual power), could not sign such a dictate that drained all content from Palestinian demands.[13]

Yasser Arafat embodied, for more than half a century, a national movement whose sole aim was to rectify the 1948 ethnic cleansing. The meanings of such rectification changed with time, as did the strategy and definitely the tactics. But the overall objective remained the same, especially since the demand for the refugees' right of return had been internationally acknowledged by the United Nations in 1948. He could not have signed the 2000 proposals and was immediately depicted as a warmonger.[14] This depiction, coupled with the provocative visit of Ariel Sharon to the Haram al-Sharif (the holy mount in Jerusalem where the al-Aqsa mosque stands), triggered the second uprising. This militarized protest represented a decade of frustration over a peace process that led nowhere—frustration thoroughly and effectively exploited by the Islamists on the ground.

Three years into the second Intifada, the peace effort resumed once more, the last time so far. The same formula was at work: an Israeli initiative catering to the Israeli public, and Israeli demands disguised as an American honest brokering. But a new component was added: the end of occupation was linked to the creation of an independent Palestinian state. This became part of the guidelines, accepted by the Israeli peace camp and even by the political center. This seemingly conceptual change within the Zionist polity has very little to do with ideological shifts in Israel, and much more with slight fluctuations in the local balance of power.

The uprising in 1987 persuaded the peace camp in Israel of the need to distinguish between discourse and realities on the ground. Hence, the discourse of a Palestinian state, once totally taboo in Israel, became now a means of selling to the Americans, and through them to the world, a solution that offers only a ministate to the Palestinians. Even a cursory examination of the details of the proposals exposes an offer not for a state, but rather a Bantustan with no independent policies, no territorial integrity, no viable economic or social infrastructure, and no capital, on just half of the West Bank, within huge walls surrounding its fragmented cantons. Only with such restrictions could the peace camp link the end of occupation with the creation of an independent Palestine. Meanwhile, the continued indifference to past chapters in the conflict—notably the 1948 ethnic cleansing—helped to sell to the Israeli peace camp the need to acknowledge an "independent" Palestine alongside Israel.

As I mentioned before, these guidelines fit a more general American outlook on peacemaking, of which the salient feature is the absence of any reference to past failures in the making of peace. Hence the obvious failures of the peace process in Palestine have never been an integral part of the thinking accompanying the next stage in the same process. The reasons for the total collapse of the Rogers and Yarnig initiatives of 1971, the Kissinger proposals in 1972, the Geneva conferences of 1974 and 1977, the 1979 Camp David initiative, the Oslo accord, and all the latest gambits are clear. They all stem from the attempt of the peacemakers to remove the ethnic cleansing in 1948 from the peace agenda and absolve Israel from its responsibility for the Nakbah. But the absence of a learning process ensures that this obvious reason would not be considered in the attempt to analyze the failure. The tendency is rather to cast the blame on the Palestinians and depict them as warmongers and intransigent people who do not wish to end the conflict. The principal victims of the conflict, whose misery and predicament have fueled the conflict since 1948, are totally absent from the peace agenda.

The reason for this bewildering absence is rooted in political psychology as much as it caused by the politics of power behind the peacemaking in Palestine. The Israeli position on the conflict and its solution is fed more by fears, psychoses, and traumas than by the security interests and concerns that are presented by the Israelis as the basis for their opening gambits in every round of negotiations.

At the heart of that fear is the knowledge that after years of denial, a peace process that would put the refugees' right of return at the focus of the reconciliation effort would open a Pandora's box. Out would come inevitable questions about the moral foundations of the Jewish State and the essence of the Zionist project. These would be accompanied by a host of questions relating to restitution rights—from financial compensation to war crime tribunals.

Israeli Historiophobia

The perception of peace is connected in Israel to the construction of Israeli national identity and the institutionalization of a particular hegemonic discourse in social and popular culture that entails the constitution of a Palestinian/Arab self as its demonized "other." It is therefore useful to focus on the implications of presenting Arab identity as the "other" of Israeli national identity for potential reconciliation in contemporary Israeli society.[15] Such a connection highlights the relationship between victimhood, justice, and the prospective solution. As I have shown elsewhere, the rootedness of this discourse of otherness and its prevalence in Israeli popular culture has posed a key obstacle to an equitable solution to the current conflict.[16]

As critical theories of nationalism have taught us, the negation of the "other" for the construction of a collective identity is central to the imposition of a hegemonic national identity. In the particular case of Israel, this formulation has taken on an added significance that was painfully exposed in the early 1950s. Beginning in the nineteenth century, but especially since the creation of the state of Israel in 1948, Arab identity has come to be constructed as the "hated other" of Israeli national identity, symbolizing everything that Jewishness was not. This juxtaposition ran into trouble when Israel encouraged about one million Arab Jews to immigrate. There was a conscious effort in the 1950s to de-Arabize these Arab immigrants: they were taught to scorn their mother tongue, reject Arab culture, and make an effort to be Europeanized.[17]

This approach to identity—that is of constructing an "other" as the negative pole of oneself—was further reinforced through Israeli historiography: specifically, as it dealt with Jewish terrorism in the Mandatory period or with Jewish atrocities in the 1948 war. Given that terrorism is a mode of behavior that Israeli "Orientalists" attribute solely to the Palestinian resistance movement, it could not be part of an analysis or description of chapters in Israel's past.[18] One way out of this conundrum was to accredit a particular political group, preferably an extremist one, with the same attributes of the enemy, thus exonerating mainstream national behavior. As such, Israeli historians and Israeli society at large have been able to admit to the April 1948 massacre in Dir Yassin, committed by the right-wing group Irgun, but have covered up or denied other massacres carried out by the Hagana—the main Jewish underground from which the future Israel Defense Forces was formed.[19]

A similar conundrum is posed by the problem of Palestinian victimhood in the context of a reconciliation process. Acknowledging the "other's" victimhood, or recognizing oneself as the victimizer of the "other" is unthinkable for most Israeli historians. For the Israelis, recognizing the Palestinians as victims of Israeli actions is deeply traumatic. To recognize the injustice involved in the death and displacement of the land's native inhabitants would not only question the very foundational myths of the state of Israel and its motto: "A state without a people for a people without a state"; but it would also raise a panoply of ethical questions with significant implications for the future of the state. In other words, this fear is deeply rooted in the Israeli perception of 1948, as well as earlier when, according to mainstream and popular Israeli historiography, early Zionists settled an empty land, making "the desert bloom." One of the founding myths of Israeli society is that of David fighting Goliath in a hostile environment.

The inability to acknowledge Palestinian trauma is also vitally connected to the Palestinian narrative of that year, as the year of the Nakbah. Had this victimhood been related to the natural and normal consequences of a long-lasting bloody conflict, Israeli fears of allowing the other side to be portrayed as a victim of the conflict would perhaps not have been so fierce. From such a perspective, both sides would have been victims of "the circumstances" or any other amorphous, noncommittal concept that absolves human beings and particularly politicians from taking responsibility. But what the Palestinians are demanding—what in fact has become a *conditio sine qua non* to many—is that the Palestinians be recognized as the victims of an Israeli evil. For Israelis to lose the status of victimhood in this instance has international implications; more critically, it harbors existential repercussions for the Israeli Jewish psyche. It implies recognizing that they have become a mirror image of their worst nightmare.

The Palestinian side has similar fears, but there is an imbalance between Palestinians and Israelis in the twin process of other-victimization and self-vilification, which I have explored elsewhere.[20] Thus, the Israeli Jewish side perceives that it has more to lose by an open exploration of history.

Educators, historians, novelists, and other cultural producers have imbedded this fear in the national narrative, in the ethos and myths of Israeli society during times of war, or warlike times. This narrative manifests itself in the tales told on Independence Day and Passover, in the curriculum and textbooks in elementary and high schools, in the ceremonies of freshmen and the graduation of officers in the army, in the historical narrative delivered in the printed and electronic media, in the speeches of politicians, in the work of artists, novelists, and poets, and in the research of academics about Israel's past and the present.[21]

Fear, therefore, plays a motivating role in the shaping of the concept of peace on the Israeli side. Victimizing the other and negating its right to the role of victim are intertwined processes of the same violence. Those who expelled Palestinians in 1948 deny the ethnic cleansing took place. And so the self-declaration of being a victim is accompanied by the fear of losing the Jew's position as the ultimate victim in modern history.

The Palestinian narrative is that of suffering, reconstructed on the basis of living memory, oral history, a continued exilic existence, and the more tangible effects such as property deeds, faded photographs, and keys to homes they can no longer return to. These narratives are read backward through the prism of contemporary hardships, in the occupied territories where residents are subjected daily to house demolitions, sudden arrests, expulsions, and more recently to daily atrocities committed by the Israeli army; and in exile where they are subjected to the whims of their host countries and in some instances denied even their most basic civic and human rights. Through this prism, Zionism or Israel has come to represent absolute evil, and the ultimate victimizer. How can this image be divided in the businesslike approach to peace preached by American and Israeli peacemakers?

It cannot, of course. When peace is discussed in such a context one should appeal to ways in which communities of suffering, worldwide, reconcile with their victimizers. The narrative of suffering is an interpretative construct describing a collective evil in the past, often employed for the political needs of a given community in the present, in order to improve its conditions in the future. In order to avoid a reductionist view of the narrative of suffering, I should add that in the case of the Palestinians especially, as well as other communities that continue to live the aftereffects of the original action, this narrative also has a redemptive value—for the communities themselves. However, and as the case of the Holocaust has shown, the way this narrative is manipulated by cultural production and political actors for political ends is another issue, one that I will not discuss here.

In most contexts, this narrative is reproduced with the help of educational and media systems, a commemorative infrastructure of museums and ceremonies, and is preserved through a variety of discourses. Even though it can serve a community in conflict, it is more difficult as means for reconciliation.

In the case of the Palestinians, living under occupation or in exile, commemoration takes on myriad and sometimes unexpected forms. Lacking the basic infrastructure, and in the absence of a terra firma on which to establish these rituals, commemoration takes form in the Occupied Territories most explicitly in crowding the calendar with significant days that have to be commemorated: days such as the Balfour Declaration, the Declaration of Independence, the End of the Mandate, the Partition Resolution and the day of Fatah's (the Palestinian Liberation Organization) foundation. In exile and often lacking the political, economic, and civic rights necessary, retelling the narrative takes on its own local color. In Lebanon, for example, where the Palestinian presence is viewed as a serious threat to the country's sectarian balance and hence long-term political stability, the mass graveyard of the Sabra and Chatila massacres—where, in the aftermath of the 1982 Israeli invasion, 2,000 residents of the camp were massacred by right-wing members of a Lebanese militia under the watchful eye and protection of the Israeli army—has been used as a massive garbage dump for the past 18 years. Every year it is cleared up in September, but it usually takes activists from outside the camp to generate some memorial event before it disintegrates into a dump again. More recently, children in these same camps have transformed the commemoration of the Nakbah through a retelling of their own personal narratives and imaginative reconstitutions of the Palestine to which they wish to return.

In another exilic community in Tunis a group of Palestinian activists transformed their private living rooms between 1983 and 1993 into live museums of the catastrophe that had befallen their people. In each living room a small corner was set up representing their families' own narrative and discourse of national identity. In yet another example, in Cambridge, Massachusetts, Palestinians and others have come together for the past three years on December 13, the anniversary of the first Intifada, to relate their own personal stories.[22]

History as Liberator

The following paradox, then, has to be resolved: the recognition that the refugee problem is at the heart of the conflict while acknowledging that for the stronger party in the conflict it is a nonstarter in any future peace negotiations. Solving the paradox necessitates coping with Israeli Jewish fears of the past. The most difficult part of Israel's encounter with history is the Jewish State's need to recognize the cardinal role it played in making the Palestinians into a community of suffering. How, though, will the Israelis be able to accept the implications of such a step? I will suggest three possible ways here, out of probably many others through which the violent element in the relationship between the two communities can be extricated.

The three ways, three realms that might offer some guidance, are civic and international law, sociological theories of retribution and restitution, and finally cultural studies, so as to articulate better the dialectical relationship between collective memories and their manipulation.

The very idea of considering the 1948 case as a legal matter is anathema to most Jews in Israel. In fact this mere suggestion would sow panic and horror among this particular community. Most frightening of all in this context would be the probable implication and inclusion of some Israeli Jews in the category of war criminals. Nevertheless, I believe that to achieve some form of actual reconciliation, this step has to be taken.

Such associations and insinuations would quite likely antagonize many Jews in Israel; it is understandable how little would be the incentive to ride that ghost train in the land of fear back to the past. Given the present imbalance of power between the Palestinians and the Israelis, in which the Israeli government effectively controls territorial access as well as all vital resources, any potential incentive to face up to this past diminishes considerably.

But it is worthwhile to find out what component of the legal process can be utilized. One promising area is that of litigation arising from the ethnic cleansing operations: an assessment of the destruction, breaking it down to its various aspects (social fabric, careers, culture, real estate, etc.). Some of these aspects can be quantified. One of the best means of approaching this quantification of suffering was offered by the Israelis and Germans in their reparation agreement, which included pensions calculated according to inflation across the years, estimation of real estate, and other aspects of individual loss. A different set of agreements translated into money, in the form of grants to the state of Israel, the collective human loss. In his writings Salman Abu Sitta has begun using this approach to estimate the real value of assets lost in the Nakbah.[23]

Another avenue to explore in this context is use of the existing public tribunals that handle such litigation and lawsuits. This should be examined together with the question of international intervention in local conflicts, in the wake of evidence on atrocities or crimes. One has to admit that this approach links the Nakbah of 1948 to the cases of ex-Yugoslavia and Rwanda, and implies a set of procedures for the rectification of past evils. Only an insignificant minority of Jews in Israel are willing to consider the conflict in such a context, but there are some and hence this should kept in mind.

The road forward on the legal plain is to look for non-retributive justice. Rwandan author Babu Ayindo, in his article "Retribution or Restoration for Rwanda" published in January 1998 in the journal *Africanews*, elaborated upon one possible strategy. Dealing with the International Criminal Tribunal for Rwanda (ICTR), Ayindo writes:

> suffice it to say that the retributive understanding of crime and justice, upon which the ICTR is founded, is discordant with the world view of many African communities. To emphasize retribution is the surest way to poison the seeds of reconciliation. If anything, retribution turns offenders into heroes, re-victimizes the victims and fertilizes the circle of violence.[24]

Ayindo is here inspired by Howard Zehr's book, *Changing Lenses*,[25] which comes out strongly against the pro-punishment judicial system. One of the questions Zehr raises, and that is picked up by Ayindo in his discussion of the Rwandan case, is relevant to our contemplation of the means by which Jews in Israel could overcome their fear of facing the past. He asks whether justice should focus on establishing guilt, or on identifying needs and obligations? In other words, can it serve as a reregulator of life where life was once disrupted? Ayindo states clearly that justice cannot be made to inflict suffering on victimizers, let alone their descendents, but must prevent suffering from continuing. This claim that Zehr considers revolutionary, explains Ayindo, is easily understood by many people in Africa as the only reasonable way of dealing with victimhood. Even if one cannot compare the genocide committed in Rwanda to the crime of 1948 Palestine and its continued aftereffects, the mechanism of reconciliation itself is relevant.

Ayindo distinguishes between two models in this context: the tribunal in Rwanda that deals only with the past, and does not enable a reconstruction of relationships there, and the Truth and Reconciliation Commission chaired by Bishop Desmond Tutu in South Africa; Ayindo prefers the latter, because it pays attention to the future. The power underlying the Truth and Reconciliation Commission, according to Ayindo, lies both in its disinclination to inflict heavy penalties, and in its insistence on discussing future relationships between different communities in South Africa. In contrast, the first model, the Rwanda tribunal, is the fastest and surest way of turning the victims into victimizers themselves.

A second way of overcoming this fear to face the past comes from the field of conflict resolution. A growing trend in the United States is victim–offender mediation, involving a face-to-face meeting between offender and victim.[26] Though obviously unsuitable for murder cases and thus not appropriate for cases of genocide as well, it is rather more adaptable to the Palestine case. The most important aspect of the procedure is the readiness of the offender to accept responsibility for the crime. Thus, the deed itself is not the focus of the process, but its consequences. The goal is restorative justice, defined as a question of what the offender can do to ease the loss and suffering of the victim. It is not a substitute for criminal proceedings; in the case of Palestine, it cannot be an alternative to actual compensation or repatriation, but a supplement to any solution. This model bears a certain resemblance to the aims of South Africa's Truth and Reconciliation Commission.

Israeli responsibility for the Nakbah, if it were to be discussed as part of the attempt to reach a permanent settlement of the conflict, would obviously not reach the international court, as did the cases of Rwanda and ex-Yugoslavia. Or at least, this is what one can assess given the way the Nakbah is perceived by governments in the United States, Canada, and Europe. These political actors have so far accepted the Israeli peace camp perspective on the conflict, as elaborated above. However, many governments in Africa and Asia still support the Palestinian viewpoint and the situation may change, allowing the Nakbah to be discussed at trials concerning Israeli war crimes in the occupied territories since the creation of the court. As long as the balance of power remains as it is now, one doubts the possibility of establishing a truth commission à la South Africa. But the demands of the Palestinian victims of 1948 will continue to dominate the peace agenda. Their outcry would continue to face the offenders. Moreover, the fear of the offender would have to be taken into account in order that the settlement of the conflict can move from the division of the visible to the restoration of the deeper, invisible layers of past evils and injustices.

The third route is educational and intellectual; hence it can only be a means toward creating an atmosphere of reconciliation, rather than the process itself. This would consist of a dialectical recognition of both communities as communities of suffering. The demand that Israel recognize its role in the Nakbah can be accompanied by a parallel request that the Palestinians show their understanding of the importance of Holocaust memory for the Jewish community in Israel. This dialectical connection has already been begun by Edward Said:

> What Israel does to the Palestinians it does against a background, not only of the long-standing Western tutelage over Palestine and Arabs . . . but also against a background of an equally long-standing and equally unfaltering anti-Semitism that in this century produced the Holocaust of the European Jews. We cannot fail to connect the horrific history of anti-Semitic massacres to the establishment of Israel; nor can we fail to understand the depth, the extent and the overpowering legacy of its suffering and despair that informed the postwar Zionist movement. But it is no less appropriate for Europeans and Americans today, who support Israel because of the wrong committed against the Jews, to realize that support for Israel has included, and still includes, support for the exile and dispossession of the Palestinian people.[27]

The universalization of Holocaust memory, the deconstruction of this memory's manipulation by Zionism and the state of Israel, and the end of Holocaust denial and indifference on the Palestinian side can lead to the mutual sympathy Said talks about.[28] However, it may take more than this to convince the Israelis to recognize their role as victimizers. From the start, the self-image of the victim has been and continues to be deeply rooted in the collective conduct of the political elite in Israel. It is seen as the source for moral, international, and world Jewish support for the state, even when this image of the righteous Israel on the one hand, and the David and Goliath myth on the other, became quite ridiculous after the 1967 war, the 1982 invasion of Lebanon, and more recently the Intifada. And yet the fear of losing the position of the victim remains closely intertwined with the fear of facing the unpleasant past and its consequences. This is further compounded by the fear of being physically eliminated as a community, consistently nourished by the political system and sub- stantiated by Arab hostility.

Israel's nuclear arsenal, its gigantic military complex, its security service octopuses, have all proved themselves useless in the face of the Intifada or the guerrilla war in south Lebanon. They are useless as means for facing a million frustrated and radical Palestinian citizens of Israel, or the local initiatives of refugees unable to contain their dismay in the face of an opportunist Palestinian authority or a crumbling PLO. None of the weapons, nor the real or imaginary fears that have been produced, can face the victim and his or her wrath. More and more victims are added daily to the Palestinian community of suffering, in the occupied territories, Israel itself, and in south Lebanon. The end of victimization, with all its political implications, the

admission of the "other" into a national discourse, and the recognition of the role of
Israel as victimizer are the only useful means of reconciliation.

Notes

1. Some of the prominent names associated with this school are: Kenneth E. Boulding,
   Conflict and Defense (New York: Harper and Row, 1963); and John W. Burton, "Resolution
   of Conflict," *International Studies Quarterly*, 16 (March 1972), 5–29. For a fuller dicussion
   of this school of thought see Saadia Touval, *The Peace Brokers; Mediators in the Arab-Israeli
   Conflict, 1948–1979* (Princeton: Princeton University Press, 1982), 3–23.
2. Noam Chomsky, *Powers and Prospects* (London: Pluto Press, 1996), 159–201.
3. Ilan Pappe, *The Making of the Arab-Israeli Conflict, 1947–1951* (London and New York:
   I.B. Tauris, 1992), 16–46.
4. Ibid.
5. See Avi Shalim, *Collusion Across the Jordan: King Abdullah, the Zionist Movement, and the
   Partition of Palestine* (Oxford: Clarendon Press, 1988), 151.
6. Nur Masalha, *The Expulsion of the Palestinians: The Concept of 'Transfer' in Zionist Political
   Thought, 1882–1948* (Washington: Institute for Palestine Studies, 1992), 93–141.
7. The language of cleansing is used openly in Plan Dalet (Plan D). An English version
   of this plan can be found in Walid Khalidi, "Plan Dalet: Master Plan for the Conquest of
   Palestine," *Journal of Palestine Studies*, 18: 69 (Autumn 1988), 4–20.
8. Pappe, *Making*, 203–243.
9. Ilan Pappe, *A History of Modern Palestine: One Land, Two Peoples* (Cambridge: Cambridge
   University Press, 2003), 200–211.
10. On the PLO's foreign relations, see Helena Cobban, *The Palestine Liberation Organization:
    People, Power and Politics* (Cambridge: Cambridge University Press, 1984), 228–235.
11. "Permanent status negotiations will commence as soon as possible, but not later than the
    beginning of the third year of the interim period, between the Government of Israel and
    the Palestinian people representatives. It is understood that these negotiations shall cover
    remaining issues, including: Jerusalem, refugees, settlements, security arrangements, bor-
    ders, relations and cooperation with other neighbors, and other issues of common inter-
    est. The two parties agree that the outcome of the permanent status negotiations should
    not be prejudiced or pre-empted by agreements reached for the interim." Full text of the
    accords can be found at www.yale.edu/lawweb/avalon/mideast/isrplo.htm
12. I have described in detail the situation in Palestine in Ilan Pappe, "Breaking the Mirror:
    Oslo and After," in Haim Gordon, ed., *Looking Back at the June 1967 War* (New York:
    Praeger, 1999), 95–112.
13. Robert Malley, "Fictions about the Failure at Camp David," *New York Review of Books*,
    August 7, 2001, 1–7.
14. Ibid.
15. Ilan Pappe, "Zionism in the Test of the Theories of Nationalism and the Historiographic
    Method," in Pinhas Ginossar and Avi Bareli eds., *Zionism: Contemporary Controversy*
    (Beersheba: Publishing House of Ben-Gurion/University of the Negev, 1996), 223–224.
16. Ilan Pappe, "Fear, Victimhood, Self and Other's Images," in Rafiq Shami, ed., *Fear in
    One's Own Country* (Zurich: Collegium Helveticum in der Samper-Sternwarte, ETH,
    2000), 65–78.
17. The best source on de-Arabization is Ellah B. Shohat, "Sephardim in Israel: Zionism from
    the Standpoint of its Jewish Victims," *Social Text* 19 (1998), 1–20.
18. "Orientalists" are those who reify "Eastern" cultures as inferior, dangerous, or mysterious.
    See Edward Said, *Orientalism* (London: Penguin Books, 1978); and see Gabriel Piterberg,

"The Nation and Its Raconteurs: Orientalism and Nationalist Historiography," *Theory and Criticism*, 6 (1995), 81–104.

19. Ilan Pappe, "Post-Zionist Critique: Part I: The Academic Debate," *Journal of Palestine Studies*, 26: 2, (Winter 1997), 29–41.
20. Pappe, "Fear, Victimhood, Self and Other's Images."
21. Pappe, "Post-Zionist Critique."
22. See Edward Said, "The Right of Return at Last," in Naseer H. Aruri, ed., *Palestinian Refugees: The Right of Return* (London: Pluto Press, 2001), 1–9.
23. Salman Abu Sitta, "The Feasability of the Right of Return," in Ghada Karmi and Eugene Cortran, eds., *The Palestinian Exodus, 1948–1988* (London: Ithaca Press, 1998), 171–196.
24. Babu Ayindo, "Retribution or Restoration for Rwanda," *Africanews* (January 1998), 10–17.
25. Howard Zehr, *Changing Lenses: A New Focus for Crime and Justice* (Waterloo, Ontario: Herald Press, 1990).
26. See John Braithwaite and Heather Strang, eds., *Restorative Justice and Civil Society* (Cambridge: Cambridge University Press, 2001).
27. Edward Said, *The Politics of Dispossession* (London: Chatto and Windus, 1994), 167.
28. Peter Novick, *The Holocaust in American Life* (Boston: Houghton Mifflin, 1999); Norman Finkelstein, *The Holocaust Industry: Reflections on the Exploitation of Jewish Suffering* (London: Verso Books, 2000).

Suggestions for Further Reading

Aruri, Naser H., *Dishonest Broker: The US Role in Israel and Palestine* (Cambridge, MA: South End Press, 2003).

Karmi, Ghada and Cotran, Eugene, eds., *The Palestinian Exodus, 1948–1998* (London: Ithaca Press, 1999).

Nimni, Ephraim, ed., *The Challenge of Post-Zionism* (London: Zed Press, 2002).

Said, Edward W., *Peace and Its Discontents: Essays on Palestine in the Middle East Process* (New York: Vintage Books, 1996).

CHAPTER 8

Nigeria: The Past in the Present

Toyin Falola

.

L
ooking back at the series of calamities that has befallen Nigeria since it obtained its independence in 1960, Nigerians have wondered why their country is not developed, in spite of its high number of educated people, abundant resources of oil, fertile land, minerals, and other products. From a country regarded to be on the verge of greatness in the 1970s, it became one of the poorest by the close of the century. Competing interpretations of Nigeria's history explain the unhappy unfolding of events: a bungled transfer of power from the British to pioneer political leaders led from one crisis to another until the fall of the First Republic in January 1966; the army that took over lasted till June, overthrown by another group of officers. As northern officers displaced the southern ones, interethnic tensions exacerbated, leading to the killing of thousands of Igbo people in the North. A civil war followed from 1967 to 1970 and a decade of military rule ensued; but the transfer of power to civilians in 1979 only led to a short-lived Second Republic, which was aborted in 1983. This was followed by long years of military rule, with coups and countercoups, and broken promises to disengage and initiate a democracy; an election that was free and fair that produced a president in 1993 was annulled by the military, leading to threats of secession; another republic was inaugurated in 1999, officially called the Fourth Republic (the Third was aborted before it took off!). But the democratic regime that followed the authoritarian ones has been unable to stem the tide of religious rivalries, interethnic tensions, the rise of ethnic militias, and religious fundamentalism.[1] Since the 1990s, a persistent call has been for the government to convene a national conference so that the more than 250 ethnic groups can review their history and negotiate their future.[2] Fearing that past history will create problems for present politics, the government has been afraid to convene such a conference. History becomes dangerous, as its narration and lessons may end in the dismantling of Nigeria.

Why does the government oppose the retelling of history? The answer is clear enough: there is no consensus on national history, just as there is no consensus on the idea of the nation. To talk about the nation may be to revive the history that will destroy the nation. A historical review need not necessarily lead to strategies to create a stronger Nigeria, to build a stable government that will take care of the basic needs of all the citizens, or to initiate a developmentalist ideology. Rather, what the historical review promises to lead to is, at worst, fragmentation or, at best, a federal system with a weakened center and greater autonomy for the component units. With a strong center gone, the autonomous units can revive the idea of secession.

The historical problems at the center of politics revolve around ethnicity, development, and the survival of Nigeria as a single country. History shapes discussions on the nature of ethnic pluralism, ethnopolitics, adoption of a federal system of government, political values, leadership, and personality. By "history," I mean both the background of contemporary issues, as well as the way this background is perceived and interpreted both by political leaders and their followers. Political debates return to past histories to argue for or against the breakup of Nigeria into three or more countries. If Nigeria is to remain undivided, how should power be distributed? What are the rights of minority groups, especially those that produce the oil that creates the revenues for the country? The North comprises a majority Muslim population, while the South has more Christians. The religious divide, rooted in history, adds to the problems. In controlling federal power, should Christians or Muslims hold the greatest stake? The discussion of present politics usually tends to reveal the power of history, the legacy of the colonial era, and the failure of past leaders.

Thus far, I have identified the issues and crises that plague contemporary Nigeria. In what follows, I will present historical information that serves as a foundation for understanding the formation of Nigeria, which emerged under the British. Subsequently, detailed examples are given of the way each major group in Nigeria, including the Islamic North, the Yoruba, and the Igbo, as well as various minority groups, all use their own versions of history to interpret the present and to justify their claims to power and resources. On the whole, the divergent, partisan versions of history complicate the capacity for Nigeria to function in a unified way and to resolve the major crises it faces.

The "Tribe" in the Modern Nation

Politics in Nigeria has been affected by ethnicity and ethnic rivalries. History writing, too, has been affected by ethnicity. Without understanding the politics of the "tribe," it is impossible to put into context the partisan nature of historical representation. The force of the premodern exerts itself on the modern. The untidy creation of "Nigeria" by the British in the first half of the twentieth century ultimately led to a "malformed" state. Nigeria is erected on a layer of many existing indigenous states and nations. The colonial government did very little to bring the myriad nations together, and the postcolonial one has failed to do the same. The three major precolonial formations were consolidated in the first half of the twentieth century, and they became transformed into the three principal regions in a federal system.[3] Precolonial formations (Yoruba, Igbo, and Hausa–Fulani) have been carried forward to the present, thus perpetuating interethnic rivalries and conflicts. The Igbo came to dominate the East, the Yoruba the West, and the Hausa–Fulani the North. The visible hand of the past in the present is not just about the historical background of Nigeria, but the reality of history in daily governance, interstate interactions, and the negotiations that individuals have to conduct with the state and its multiple agencies.

The system of colonial administration installed by the British, known as "Indirect Rule," created its own problems. Indirect rule consolidated "tribalism," promoting various identities within the country. Although presented as a form of local government, indirect rule was premised on the difficulty of reconciling two cultures: the "superior" British one with those of "inferior" Africans. Indirect rule would allow Nigerians to be governed by the very institutions they created. In implementing indirect rule, the Sokoto Caliphate in the North, a theocracy created by a jihad (1804–1817), was chosen as the model. Considering this to be the best and most sophisticated approach to indirect rule, the British tried to extend it to other parts of the country, even among the Igbo of the East, where there had been no kings. Indirect rule reinforced parochialism at a subregional level, with people thinking more in terms of the locality than the nation.

The colonial government, having chosen to pursue a "tribalist" policy, took steps to keep various parts of the country apart from one another. Some of the tendencies of those years have been retained till today. The control of land, especially its allocation to strangers, was one such powerful tool. In colonial northern Nigeria, land was vested in the government and free holding was prevented. This made it very hard for southerners to have free access to land. "Strangers' quarters" were created for southerners and Christians; to this day in many parts of the North, they live separately from indigenes and Muslims. Thus, in cases of violence, and these have been common, it is easy to identify the strangers and destroy their properties and kill them. In the South, where traditional land tenure practices were retained and land commercialization began early enough, there were also separate quarters for northerners, making integration within cities and regions very difficult to attain. In all cases where strangers and citizens live apart, the construction of identities differs, and historical narratives on the same event also vary. Notions of "citizens" and "strangers" have emerged into ideas of separate identities, different histories. If citizens talk about the spirit of accommodation, strangers point to that of victimization.

Regionalism, and the separatism that came with it, was slowly entrenched,[4] even acquiring the force of law, rationalized by traditions and myths. The North and the South were kept apart as much as possible; citizens living outside their areas suffered discrimination; and uneven development between the regions fuelled bitter politics. By the late 1930s, British officers and some Nigerians accepted the reality of the "tribe," forgetting that they themselves had contributed to its "invention" in the preceding years. They now decided to reorganize the country on the basis of "natural divisions." By "natural" they were not referring to rivers, lakes, and mountains, but constructed histories and traditions that divide people. Rather than "invent" a nation, they would manipulate the "divisions." Thus in 1939, the southern protectorate was broken into two, the western and eastern groups of provinces, paving the way for the creation of regional assemblies in 1946.

As the country moved toward independence from the 1940s onward, the British, in collaboration with Nigerian political leaders, began to take steps to consolidate tribal/regional politics. Controversies began to emerge over the constitution to administer the regions, as many advocated autonomous powers to the three regions of East, West, and North. Any arrangement that appeared to be "unitarist" was rejected in preference of one that promoted regionalism and federalism. Three major regional political parties emerged to create the basis of "one-party regions" that reinforced the divisions and promoted the politics of ethnicity. These parties were the Action Group among the Yoruba in the West, the National Council of Nigeria and the Cameroons (later the National Council of Nigerian Citizens) in the East, and the Northern Peoples Congress in the North.

At a contentious national conference in 1950, the North made its demand to be part of Nigeria very clear: it would not lose any of its territory to other regions, as the West was asking for a kind of merger with some areas; its number of seats in the central House of Representation would be 50 percent; and population had to be the main criterion to distribute nationally collected revenues, thus diverting the greater share of resources to its region. The East and West were not happy with some of these demands and the belief that the North was trying to dominate them gained ground. By 1953, another conference had to be convened in London, after the North threatened secession. The adoption of a federal constitution followed in 1954, to formalize the separatist tendencies already under way. Federalism accepted the "tribal" basis of Nigeria, and the uneven political and economic development of different regions. Yet the force of history is potent: the problems that led to the adoption of federalism refused to disappear. If the country arrived at this decision because the North was alleging domination, it was now the turn of the South to make the same claim. The federation became dominated by the huge North, whose population and size were more than all the South combined. The North's representation in the House of Representatives was originally 50 percent, later 52, and it controlled executive power.

If the previous history and experience of divisions led to federalism, the post-1954 arrangement created a new history that would become the justification for the chaos and wars of the 1960s—not to mention the fall of the republic within five years of independence. There was no agreement as to when the three regions should become self-governing, with the West and East opting for self-government in 1957 and the North in 1959. With its numerical dominance, the North was able to produce the first (and only) prime minister, Tafawa Balewa, in 1957. It was a great act of compromise for the three regions and parties to form a government on the eve of independence in 1960. The regions began to compete as well as to articulate and defend their interests. Each had its judiciary, public service, marketing board, and other agencies. Recruitment to these agencies was based on regional citizenship, and discrimination was regarded as legitimate. In seeking investments and recognition, each region tried to establish independent consulates abroad. The country moved from a colonial status to an independent one with a full baggage of problems. Even the analysis of the problems has been so contentious that the British blame the Nigerians, the Nigerians blame the British, and one ethnic group blames the others. No party wants to take responsibility or admit to its "sins."

The Relevance of the Past: Managing a "Troubled" Legacy

There can be no doubt that elements of colonial history have shaped postcolonial and contemporary Nigerian politics and society, providing us with the data to analyze actual history and its ethnic interpretation. The problems of economic development and political management have their roots in this history. The recent past is the colonial period, which produced contested identities and an economy that is not working. The question is the extent to which the colonial past has damaged the country's present. The failure to overcome the legacy of the past has exposed the contradictions in the system, as well as the failure of postcolonial leadership. The contradictions generate multiple historical interpretations about the present.

The first history that is invoked is usually that of the "tribe." Modern Nigeria is a tripod of the three dominant "tribes," and, as noted earlier, the formation of political parties has reflected this regional division. A pattern of "bloc voting," whereby an ethnic group solidly votes for one party, shows the extent to which ethnicity influences public opinion. The country inherited a federal system with various anomalies: the problems posed by the big North, the failure to meet the demands of minority groups for autonomy and political representation, and the excessive power enjoyed by the regions at the expense of the center.

The postcolonial state has been consumed by discussions and debates on the historical relevance of ethnicity, and the choice between political centralization and regionalism. Democracy has proved to be a liability in dealing with the crises of history. All ethnic groups complain that they do not receive adequate resources from the federal government. All the leading politicians of ethnic groups want to hold important political offices, not temporarily, but for long periods. Thus, "politics" can mean what power and resources a person or an ethnic group can obtain. Political formations reveal and reflect the ethnicity shaped by past histories. And as political actors compete, they invoke past histories and present identities to support their cause.

The second history is related to the concept of the "tribe": members of the political class think and govern in terms of regionalism or ethnicity, a divisive notion that defines citizenship strictly by place of birth. The argument that ethnicity is the best way to organize society is a tenacious one, in spite of the objection by the Left that it should be class. Access to public goods and opportunities is open only to the members of a particular region or state, and nonmembers living among them suffer discrimination. Nonmembers are expected to exist at the margins, occupying the space and land allotted to them. Federal institutions that are expected to be open to all are also victims of regionalism, as entry to them can be tied to the state of origin. Competition becomes vigorous, shaped not by merit or performance, but regionalism. The regional structure that emerged in the colonial era has become the "regional ideology" of contemporary politics. Ignoring some coordination and cost-saving devices, the regions have competed in establishing economic projects. They all wanted to expand access to education, and have created many new schools, including universities. They often pursued different foreign policy agendas, with the West showing sympathy for the Soviet Union and Israel, and the North befriending Islamic countries, notably Saudi Arabia.

A third historical legacy for the postcolonial era is economic. Indeed, Nigeria has yet to change the economic structure established during the colonial era. Thus, we have the economic system of the past still dominant in the present. The need to change the structure has been a dominant historical discussion and source of debate: how can Nigeria reduce dependence on the West? Why does the country rely on a limited number of goods for export to sustain itself? Can there be a sustainable economy in an atmosphere of political instability? The answers always require a return to history in ways that either blame or exonerate the colonial period. Traditional economies began a process of modernization and greater integration with the world market under colonial rule. A new infrastructure of railways and roads emerged to move goods from various parts of the country to the ports. The network remained, integrating the country, in order to service both domestic needs and international markets. A new economic structure also emerged: the extraction of minerals and raw materials for export. Today, Nigeria can be described both as a "rentier state" and an "extractionist state." The function of the state is simply to ensure that peasants and companies "extract" materials from the land and sea (e.g., peanuts, oil, cocoa) and sell them abroad. The state collects revenues in the form of customs dues and taxes. To collect its revenues, it has to maintain order and security, using the army and the police to maintain the continuity of extractions. Since the products are not internally consumed, Nigeria becomes a peripheral outpost of a global capitalist system. The country's role is to ensure that extraction continues, and for its political class to cooperate in the exploitation of its people. Still in line with the colonial pattern, the country has to import, both essentials and luxuries, to satisfy the increasing demand for foreign goods. The colonial economy did not support the promotion of manufacturing and industrialization, and the postcolonial has been less vigorous in doing this since revenues from oil are easy to collect as "rents."[5]

In a bottom-up perspective, the historical narrative is that no one cares about the poor. The state is "irrelevant"—the colonial state did not bother with the welfare and progress of its citizens; and the postcolonial has minimally pursued the agenda of development. The colonial state handed over the management of the economy to private firms, mainly external. Caring only about places that generate resources, the firms and the government ignore the needs of the majority of the population, especially those in rural areas. A history of state irresponsibility radicalizes civil society, pushing the people to take to violence, to distrust politicians, and to question the relevance of democracy as a system of government. As one informant cynically told me in 2000, the "demo" in democracy is a "demon" that is crazy! Civil society has to fight a "mad demon" with a power superior to madness. The history of economic failures has instigated violent demands in contemporary society, linking economy and politics in clear ways. To civil society, Nigerian history "from below" is that of a series of oppressive events and resistance: actions and praxis merge to generate "radical history." The religious-cum-political riots in the 1980s, and the antimilitary resistance and the pro-democracy movements of the 1990s are examples of political actions based on a different reading of contemporary history.

The fourth history that shapes the contemporary reality is the nature of external contacts. There were two forces, each with its own distinctive stamp. The first and older influence is Islamic, from the Middle East and North Africa, via the Sahara desert. Long established trans-Saharan trade brought goods, Islam, and various aspects of Arabic civilization to the areas of modern northern Nigeria. Over several centuries from the eleventh century onward, Islamic and Arabic civilizations spread to various parts of the North. Areas where Islam was accepted became Islamic, and their rulers exchanged contacts within the region and with some of their colleagues in North Africa and the Mediterranean. During the nineteenth century, a jihad led by Uthman dan Fodio created the largest theocracy in Africa. The presentation of history in the North privileges Islam and contacts with the Islamic world.

The second influence is from Europe, via the Atlantic Ocean. Western education and Christianity spread in the South from the mid-nineteenth century onward. By the colonial period, the modern intelligentsia was dominated by southerners. Whenever "progress" and "civilization" are defined in Western ways, southerners claim to be superior, even till today. Southerners have always insisted that the power of knowledge should translate to political and economic power. For southerners, history is about Western-style modernization and the obstacles to it.

Both influences solidify the divisions in the country. Northern political leaders, doubling as an Islamic intelligentsia, regard the Islamic North as superior in civilization to the "pagan" or "Christian" South. The British actually agreed with them, regarding the intellectual, political, and economic achievements of the Sokoto Caliphate as superior to those of the South. The belief created imbalance in education and social development for many decades. While the British were entrenched in colonial Nigeria, they believed that the colonial administration would be stable and the Islamic leaders more prone to peace if they controlled the spread of Western influences. A so-called pact signed in the first decade of the twentieth century between northern emirs and the governor prevented the coming of missionaries to establish churches and Western schools. Since the missionaries were initially the agents of Western education, their attention concentrated in the South. Historical narratives from the South tend to focus on the West as an agency of change.

The educational imbalance is yet to be corrected, and is now an integral part of political competition. The statistics tell a small part of the story: initially all schools were located in the South; in 1912, the South had 150 primary schools and the North had 34; and when the South was demanding more grammar schools, the first one in the North was established in 1922. On the eve of the country's independence, the gap had widened: the South had a university; over 15,000 primary schools with almost 3 million pupils, and over 180 grammar schools with about 30,000 students. The North did not have a university; there were slightly over 2,000 primary schools with about 200,000 students, and only 18 secondary schools with about 4,000 students.

Since the colonial era, southerners continue to maintain that northerners have slowed down the country. The way southerners have presented the argument has shifted from era to era. During the late colonial era, southerners argued that the British should transfer power to them because they were the most educationally advanced while northerners lacked a notion of modern development. As independence approached, southerners turned their head start into a political argument, connecting their possession of Western education with a bid for the control of modern politics. While the North had produced only one person with a university degree by 1950, the South pointed to hundreds and their so-called capacity to change the country for good within a short time. Northerners have always dismissed the argument, saying that Islam had offered them a basis for intellectual development, that their literacy level was the highest in the country, and that they did not want the kind of Western-oriented development that southerners advocated.

In the early years of independence, the arguments were repeated. Northerners accused the South of dominating the civil service, police, and other modern sectors.

As far as northerners were concerned, the inroads of southerners to their areas should be curtailed; meanwhile, the North accelerated the development of Western education to produce its own educated citizens. As the British transferred federal power to northerners, southerners saw them as unfit to govern, and accused northern leaders of lacking the skills and ideas to manage a modern nation in a hurry. The northern leaders fought back, accusing their southern counterparts of schemes to dominate them. Accusations and counteraccusations of domination have since become so rampant that practices for hiring and contracts are based on what is known as "federal character," which is to represent all regions in the distribution of money and positions. To the South, "federal character" puts "tribalism" above "merit" and curtails the rise of southerners.

Since northerners have controlled federal power for the greater duration of the postindependence era, southerners tend to blame them for causing underdevelopment. Northerners always respond that economic and bureaucratic power resides in the South, due mainly to its educational advantage. The military wing of the northern political class has depended on violence and military coups to acquire power. The "intellectual" wing of the same class has argued that a three-tier privilege should emerge: the Igbo in the East should control bureaucratic power, the Yoruba in the West should have economic power, while they in the North should dominate politics. One does not have to be a leftist to accept the reality that power is at the center of economics and bureaucracy, and that of the three elements identified, the control of politics is the most lucrative. Whoever controls federal politics allocates oil licenses, the key means to wealth. According to the southerners, northern politicians want southerners to use their educational skills to produce the oil, while they sell it to make the money. Thus today, the history of past progress in relation to the attainment of Western education continues to have an impact upon the nature of political competition, the vigorous interregional conflicts over jobs and contracts, and the allocation of federal grants to various parts.

In what follows, I will explore in brief the four major themes that dominate contemporary political discourse and influence many political activities and decisions.

North: The Politics of the Jihad and Islam

The jihad (1804–1817) is the central theme in the historical narrative, the subtext of modern politics. Without understanding the jihad, one cannot analyze the nature of contemporary politics. The jihad created a large theocracy, the Sokoto Caliphate, which brought under one political umbrella various Hausa kingdoms (now emirates), including a number of non-Hausa ones in the Middle Belt, and further down to include the Yoruba state of Ilorin. Power became hierarchical and centralized, with emirs (kings) accepting the authority of the Caliph based in Sokoto, the religious and political capital city. The Caliph became both a spiritual and political leader of a large political unit covering the large part of what we now call northern Nigeria. Although the Caliphate had entered, by the turn of the twentieth century, a period of moral and political weakness, the British empowered it, strengthening the power of the kings, the prestige of Islam, and the ideology of a centralized state. The history of the Caliphate has become embedded in modern politics—its interpretation and values are inscribed into contemporary modernity, and the ideas of dan Fodio are worshipped.[6] Some of the influences of the jihad are pervasive; even in managing modern economies, some people refer to the statements and ideas of jihadists.

First, the basis of a united North, or a belief in the political concept of "One North" advocated by the northern political class, is based on the legacy of the jihad. Islam has become not just a religion but also a political agency. Power becomes vested in Muslims, with non-Muslims not only excluded, but also not expected to hold or contest power. Indirect rule, mentioned above, consolidated the power of the Sokoto Caliphate, in ways far stronger than the British intended. The history of the jihad becomes a way to legitimize power for those who have it, or to attack the credentials for power of those who lack it.

Second, the majority groups have turned to Islam and the legacy of the jihad to maintain political dominance on the minorities living among them. The North does not just comprise the two dominant ethnic groups of the Hausa and the Fulani. However, the jihad and Islam not only gave them the opportunity to consolidate their identity, but also to attempt to undermine the interests of minorities. As the Hausa–Fulani majority tried to dominate, the minorities tried to rebel, leading to many cases of riots, violence, and complicated religious tensions. In the "ideology" of the jihad, non-Muslims (called pagans) should not have power over Muslims. If they cannot be converted, they can be exploited to pay tributes and more taxes. By using Islam, the Hausa–Fulani turned themselves into the center of power, their areas as the core. Non-Muslims and non-Hausa–Fulani were constructed as "inferiors." Powerful kingdoms such as the Nupe and Igala, and proud groups as the Jukun and Tiv were, over time, "peripheralized." The agenda in the nineteenth century was to use force to convert them. Many resisted the jihad's attempts at forced conversion, and used indige-nous religion as a powerful repellant to Islam. Resistance to Islam is a major historical theme among the non-Hausa–Fulani, and to narrate their experiences has become a way of affirming their ethnic solidarity, creating and presenting their own heroes.

Third, during the colonial period, many of the minority groups turned to Christianity, as an oppositional religion to Islam with its own minority identity. Christianity has spread slowly, but maintaining peaceful relations with Islam has been a recurrent problem. Ethnic tensions may take the form of Christian–Muslim antagonism, with mosques and churches set on fire. The colonial government tried to limit the influence of the missionaries, preventing them from reaching the "core north." The missionaries chose the minority areas, and Christianity spread in the Middle Belt. Islamic leaders responded. Unlike during the nineteenth century, the use of war and violence became difficult. Instead, an Islamic missionary enterprise was embarked upon to convert "pagans" and Christians, while the use of the Shari'a enhanced the value of Islamic laws at many local levels.[7] The power of "indirect rule" manifested itself fully, as the British, in collaboration with Islamic leaders, appointed the agents of the Caliph as the political leaders of the minority groups. Part of the postindependence crisis is how the minorities have expelled political leaders who are Muslims.[8] History becomes one of the projects to attain it; the revision of the past without its Islamic elements.

When the first major political party emerged in the North in the 1950s, it acquired its power and popularity on the basis of Islam and the Caliphate. The Northern Peoples Congress (NPC) was proud of its conservative credentials, while its agenda was clear: the preservation of the power and privileges of the *Masu Sarauta*, the group that had been in control of politics since the jihad.[9] The NPC did not intend to give voting rights to the poor so that they could remove traditional chiefs and emirs from power. Politics in the North has also been driven by the desire to address the fear that southerners will dominate them. Sir Ahmadu Bello, the Sardauna of Sokoto, emerged as the dominant political figure. He was unapologetic about Islam and the jihad, and the connection of both to modern politics. His argument was that if the British had not come to stop his ancestors, the jihad would have spread all over the South, turning Nigeria into one huge Islamic Caliphate.[10] Some even see the bigger impact of Islam and Hausa culture in the region, with the thesis that the British prevented the emergence of its hegemonic power.[11] The Sardauna encouraged the spread of the Shari'a Islamic law, missionary activities, pilgrimage, and the establishment of mosques and Quranic schools. As far as he was concerned, the country should be Islamic. Bello's desire to convert many Nigerians to Islam has become the subject of partisan history, the retelling of the past as episodes of the success of Islam over "paganism" and of ongoing ambitions.

The Sardauna was not simply presenting the "devout" face of Islam for all to see. Rather, what he did was very much connected with politics—his goal was to use history and politics to create a "monolithic united entity" in the North. Bello himself had narrowly missed becoming the Caliph; his title of Sardauna was one of the most influential in the caliphate hierarchy. The Caliphate was based on a view that Islam was the source of power and legitimacy. A ruling class emerged, transferring power from one generation to another, on the basis of the legacy of the jihad. The Sardauna extended the legacy to the modern era. The NPC carefully connected political awareness to Islamic propagation, manipulated religious institutions for political ends, and drew on the unifying elements of the jihad to invent the idea of oneness, with a slogan, "One North, One God."[12] With all the forces it could muster, the NPC suppressed alternative parties, especially radical ones, even when their leaders were Muslims.[13] On the basis of this relationship between Islam and politics, the NPC was able to exclude not just southerners from joining them, but even the Christian minorities in the same region. The past—based on Islam and a history of the jihad—has become the most important tool. The Sardauna himself summarized the historical connections between the present and the past, describing the modern North as an extension of the old:

> It [self-government] can be called the restoration of the pre-1900 era, modernised, polished, democratised, refined, but not out of recognition; reconstructed, but still within the same framework and on the same foundations; comprehensible by all and appreciated by all. The train, the car, the lorry, the aeroplane, the telephone, the hospital, the dispensary, the school, the college, the fertiliser, the hypodermic syringe have all transformed Othman dan Fodio's world, but the basis is still there. The old loyalties, the old decencies, the old beliefs still hold the people of this varied Region together.[14]

The effective use of past histories for present politics survived the Sardauna, who lost his life in the coup of 1966. He has become the modern hero of the North, with a university, scores of roads, teaching hospitals, and buildings named after him. He has become the political model that others aspire to become. The history of the jihad and that of the Sardauna have merged, in grandiose proportion as "official history," to ensure that the North is held together as a solid political bloc. If the jihad created a large area unified by Islam, the Sardauna used that history to consolidate the Hausa–Fulani political hegemony. His successors have turned to the power of history to maintain that dominance.[15] The interests of the political class are articulated by the media and political parties.[16] The use of Islam has generated problems of religious conflicts,[17] with an impact on how the history of Islam and Christianity is presented in a competitive manner.[18] The Shari'a, as Islamic law, together with other aspects of Islam, is turned into a political platform, all in order to retain power. The history of the North is deliberately presented as that of the growth and impact of Islam, all in order to contest the secularity of Nigeria and empower the political class. Thus far, the members of the northern political class have been hugely successful, a success that has turned "partisan history" into "official history."

West: The "Children of Oduduwa" and the Politics of Ethnicity

As the Hausa–Fulani use Islam, other groups turn to ethnic histories to solidify and manipulate their identities in the contemporary world. Ethnic history is a common example of "partisan history," one in which a people turns the past into a source of identity and politics. Elements of the past can be invented to meet the demands of present politics. The Yoruba represent the best example of the use of ethnicity and ethnic history in modern politics. Numbering over 20 million, they have inhabited the area of western Nigeria since ancient times. Astute political leaders among them imagined a "useful" Yoruba to build a powerful identity and negotiate with others. In the classic story and format of ethnicity, the Yoruba share a common ancestry, language, and a set of religious and social beliefs. They trace their history to a single father, Oduduwa. In addition, they have a town of origin, Ile-Ife (now the city of their major university). It was from Ile-Ife that the "children" of Oduduwa spread to other parts of Yorubaland, establishing various dynasties and cities.[19]

The origin myth became the first basis in the creation of a modern political party. Chief Obafemi Awolowo, the first premier of the Western Region and the dominant hero of the Yoruba during the twentieth century,[20] created the first mass party on the basis of the Oduduwa story. He began by creating a cultural association, the Egbe Omo Oduduwa, which promoted the idea of Yoruba unity and the retention of many traditional practices. The originality of Awolowo was to tap into history to create a cultural project, and to immediately convert a political project into a political party. For the first time in Nigeria, history and tradition converged to create an identity in the service of modern politics. This project has been sustained to this very day. Awolowo was cleverly building on a strong foundation of Yoruba ethnicity. The idea of a "Yoruba nation" and Yoruba "consciousness" gradually became stronger from the mid-nineteenth century onward. When Yoruba acquired a written form during the nineteenth century, one of the achievements was to use the language to build Yoruba nationalism. Literacy allowed the Oduduwa myth to spread.

Similarly, literacy enabled the second criteria of ethnicity to gain dominance: the use of a common language (with various dialects) to express the belief in common cultures. The legendary Oduduwa and the Yoruba language combined to create a powerful ethnic group. An emergent elite constructed the idea of a stronger Yoruba nation. Although various Yoruba groups and city-states engaged in prolonged wars during the nineteenth century, the educated elite tried to disregard the war in advocating the building of an ethnic group. Through the use of a standard Yoruba language, the elite began a project of presenting cultures and institutions as homogeneous, so that one can talk of Yoruba politics, economy, and cuisine, among other aspects. A "homogeneous" culture becomes a powerful tool to fashion an ethnicity.

An ethnic identity can translate into power, prestige, and opportunities. During the colonial era, the Western-educated elite moved into various formal sector economies, worked as agents of the church, became professionals as medical doctors, attorneys, surveyors, and so on. The Western-educated elite pressured the government to create more schools, including a university. They worked hard in their churches and towns to start new elementary and high schools. The schools and their products became part of the idea of "progress," the copying of some Western ideas and their institutions. As they called for more reforms or demanded evidence of progress, the ethnic group was appropriating developmentalist ideologies. Then and now, the ethnic group measures success by those things that it has in greater excess than others. Universities, roads, and industries become part of the tools of competition in the country.

Those who seek power beyond the local government level have had to fall on the concept of Yoruba unity. Doing so means an affirmation of ethnic history, presented in many books, pamphlets, and newspaper essays. It also means the presentation of a history of ethnic relations, in ways that present the Yoruba as a dominant group, intellectual leaders, and champions of modernization. A narrative of progress as ethnic history becomes a project in the service of politics.[21] All the leading Yoruba leaders who had aspired to the presidency of the country—Obafemi Awolowo, Lateef Jakande, Bola Ige, and M.K.O. Abiola—have had to repeat the ethnic history around Oduduwa to unite their political base.

East: The Politics of Genocide and War Trauma

The Igbo is the third ethnic group in terms of dominance. Like the Yoruba and Hausa–Fulani, their leaders manipulate ethnic history for political purposes. However, unlike their rivals, they have the memory of suffering to recall in the competitive political space. They suffered the most from the first coup of 1966, and they initiated a secession that they lost. The war of secession, known as the Biafran War, lasted from 1967 to 1970, and consumed what is estimated as one million lives. In most other wars, it is the victorious party who writes the accounts of war to explain its success, to portray its triumphalism. In the case of Nigeria, it is the Igbo, who lost the war, who have been writing the most about it, turning the secession (a failed political move) into a history of victimization. The narrative of victimization is the leading political discourse in Eastern Nigeria.

Ethnic tensions came first, and still continue. Located in the east, the Igbo number more than other groups in the region. Their precolonial formation comprised hundreds of villages and towns, each autonomously governed.[22] Like other groups, the Igbo were united on the basis of trade with one another, political interactions (including wars), formation of secret societies and age–grade associations that drew members from various places, and subscription to a set of worldviews.

The British imposed a system of indirect rule on the Igbo, choosing warrant chiefs with greater power than the Igbo associated with the title. Western education spread, and the Igbo became predominantly Christian. Although unlike the Yoruba or the Hausa–Fulani, the Igbo did not evolve large centralized kingdoms, ideas of pan-Igbo identity spread during the colonial era. Chief Nnamdi Azikiwe (popularly known as Zik) emerged among them as the towering political figure. Zik started with a bigger vision of pan-Africanism, then one of the earliest to espouse ideas of Nigerian nationalism in the 1940s, but became an Igbo leader from the 1950s onward.[23] He contributed to the invention of a strong Igbo identity and nationalism, now the major theme of ethnic history.

Azikiwe's career reveals another lingering tension in Nigerian history—the development of nationalism and the idea of the nation at different times between the South and the North. Azikiwe was not the first "nationalist," but he represented a watershed in the anticolonial development that started in Lagos and left the North out of it. By the first decade of the twentieth century, political associations had emerged in Lagos, and by the 1920s, a political party, the Nigerian National Democratic Party, had been formed. An articulate media had also been created. The first party to be formed in the North was in the 1940s, and their newspaper, established in 1935, also came much later than parties of the South. By the time Azikiwe emerged on the scene, there was a political tradition in place. He energized the nationalist movement, drawing more followers. But he also began to assume the mantle of leadership for a possible pan-Nigerian political party. The problems of ethnicity set in, with the Yoruba forming their own political party and the northerners never accepting him. He had subsequently to consolidate his Igbo credentials.

The South has failed to be as politically powerful as the North in part because of a never-ending rivalry between the Igbo and the Yoruba. Here is one of the best cases of competing historical interpretations in the history of Nigeria. To both, a pan-Nigerian nationalism was developing until the 1940s when the rivalry created a split. Each group blames the other for bringing about this rift. Awolowo and Azikiwe became the leaders of the opposing groups, and each undermined the other whenever it was necessary to do so. Indeed, they both competed in presidential elections, in ways that prevented the emergence of any southern solidarity.

In consolidating an ethnic consciousness, the Igbo identity emerged as a political force. Like the Yoruba, Igbo leaders presented their group as a modernizing force for the country. The self-presentation as modernizers, still an ongoing project, added to the interethnic crises, whose escalation was to cost thousands of Igbo their lives. As the Igbo obtained jobs in the formal sectors of the economy (notably federal agencies such as the railways and post offices, and also migrated to the North), resentment gradually built up. Statements by Igbo leaders, perceived as arrogant by northern leaders, added to the problem. In spite of these resentments, recent Igbo contenders for high-stake offices such as Odumegwu Ojukwu and Alex Ekwueme continue to say that the Igbo are the modernizing agents to transform Nigeria.

Igbo historical narratives are driven by the civil war and its aftermath. The prewar years are treated as ones of persecution, and the postwar years as those of marginalization. The "theory" behind the data is clear and simple: the Igbo are great achievers, envied by other groups who do not want them to make additional progress. There is "messianism" in the story line: if other Nigerian groups cooperate with them, they have the skills and talents to develop the entire country, in all aspects, from the economic to the technological. The story line itself is that after the civil war, the Igbo have been abandoned by the federal government—the reconciliation and reconstruction that they were promised have not been fulfilled. A catalogue of woes is presented: deliberate "acts of vendetta,"[24] a high rate of joblessness, financial losses, the failure to recover "abandoned houses" in some parts of the country, inadequate representation in federal positions, and many others.[25]

Minorities: The Discourse of Marginalization

The three regional-cum-ethnic tripod legs on which Nigeria is based hide the existence of countless other groups and may even disguise the profound tensions in each region. A hegemonic group, in claiming control of an area, tries to erase the existence of others or undermine their importance. In the North, where minorities are also non-Muslims, the struggles have been to reject a pro-Islamic leadership. It is possible to assume that if the majority ethnic groups did not develop regional nationalism and establish political dominance, the minorities, too, would not have developed their particularistic identities. However, as the minorities complained about discrimination, limited opportunities, lack of development in their areas, cultural domination and exclusion from power, they began to rewrite the history of intergroup relations and to adopt strategies of separatism and violence. Even their music, stories, and oral traditions are counterhegemonic.[26]

The ideologies that shape the presentation of history include those of marginalization, of presence only as footnotes to the history of others, and as victims of Nigerian politics. Established historical traditions have been revised to minimize the impact of past histories of wars of conquest and subordination to other groups. History can even be falsified in the attempt to attain power in contemporary politics. With every demand for adjustments to boundaries or for the creation of states, "new histories" are told to justify requests and meet specific ends. The chair of a powerful government panel had this to remark in 1975:

> We were to discover that in some areas, history was distorted by people who were widely recognized as belonging to the same linguistic stock or ethnic group to convince us that they are not ethnically the same in their bid to get a state of their own.[27]

The distortion of history is a widespread practice when groups and people are eager to make demands for autonomy and development. Ethnic histories may clash with national histories in the attempt to defend specific causes. Since ethnic leaders see the country more as an agglomeration of ethnic nationalities, more emphasis can be placed on particular histories that produce various alternatives and counter-discourses.

Minority ethnic groups have developed their own particular identities and nationalism, constructed around the history of power struggles, and the memory of marginalization. Today, minority groups look back into history and see suffering at the hands of the dominant ones. Those in the oil-producing region of the Niger Delta complain that their oil revenues go to develop other areas. Where they complain against the government and powerful oil industries, they see persecution. Political narratives, as in the case of the Ogoni who have lost their leaders (notably author and political leader

Ken Saro–Wiwa) to state execution, emphasize their "punishment" at the hands of a government controlled by major ethnic groups. They also call for reparations to compensate for past losses, environmental damages, and underdevelopment. The vision of national identity portrayed by the minority groups differs from those of the majority ones—if the majority stress the need for the unity of the country, the minorities see secession or autonomy as one way out. In presenting their histories, the minorities are not shy to claim that they are being forced into a union where they exist marginally as "slaves."

The minority groups saw the handwriting on the walls quite early. As the dominant groups and the political parties that represented them consolidated the politics of regionalism, they marginalized the interests of minorities. The minorities, too, turned to the power of ethnicity, using history and traditions to solidify their identities. As with the dominant groups, they too established their own political parties.[28] The minority leaders were astute enough to realize that the dominant parties would exclude them from regional and federal powers. Consequently, they refocused the nature of political debate to the creation of new states, one that would create political units for them. While the political parties and interest groups mushroomed, the leading pioneers include the United Middle Belt Congress, the United National Independent Party, the Niger Delta Congress, the Benin Delta Peoples' Party, the Mid-West State Movement, and the Bornu Youth Movement.

The parties and their leaders have followed certain patterns, many of which remain current. To start with, they realize the futility of working within the dominant parties. Thus their own parties give them more power to bargain. Second, ambitious politicians among them gain more leverage as leaders in their own right rather than marginalized members in the dominant parties. Third, rather than play the politics of incorporation, they pursued that of opposition. Thus, each region has its "enemies from within," usually a minority party opposed to the dominant one. Finally, they are strong defenders of federalism and state creation, based on the need to become autonomous. It is possible to see state creation as a solution because the minorities constitute distinct ethnic groups with clearly defined territorial boundaries, histories of long habitation in their regions, and previous experiences of autonomy.

The majority–minority struggles became intense in the 1950s when the dominant regional parties attacked the minority ones and sought a weaker federation and regional autonomy. The dominant parties got their way encoded in the federal constitution of 1954. The minorities complained of unfairness but were ignored. Their leaders and political leaders opted for a strategy of alliance with the dominant parties, in the hope that they would obtain rewards. As it turned out, they had to reach out to parties outside their own regions, thereby complicating regional politics. In the Middle Belt, the Tiv took to violence to press home their points, initiating a strategy that remains effective up until now. As a preface to what would later become a deluge, an inquiry known as the Willink Commission was set up in 1956, to "enquire into the fears of the minorities and the means of allaying them." The various memoranda submitted to the commission are valuable historical documents on the discourse of marginalization, trauma, and punishment. The commission agreed that minorities had suffered discrimination, and had serious fears. However, the commission disagreed that there was a need to create new states for them, arguing instead that the resolution of the environmental crisis in the Niger Delta and a bill of rights to take care of discrimination would solve the problems.

The conclusions of the Willink Commission did not withstand the force of history. The environmental crisis in the delta continued. The domination led to more violence and the subsequent creation of states. From three regions, Nigeria subsequently became divided into 36 states, thus allowing the minorities to govern themselves. Two issues remain, both creating distrust and ill-feeling that fuel historical interpretations and revisions. One has been the distribution of power at the federal level, where minority groups complain that they do not have enough representation. The history of Nigeria, as they narrate it, becomes that of exclusion from federal power. Indeed, a perspective argues that minorities benefit more from military regimes than democratic ones, and a history of oppression can become one opposed to democracy.[29] The other is the distribution of oil revenues. The country's extensive oil reserves are located among the minorities in the Niger Delta. They argue that the bulk of the revenues go elsewhere, controlled by the dominant ethnic groups, and that their environments are damaged to serve the interest of the country. The fight over oil revenues, usually presented as the scheme of the majority to dominate the minorities, provides the context to understand contemporary conflicts, violence, and the presentation of history.[30]

Conclusion

Competing histories, varying visions of national identity and ethnicity, and traumatic memories of war have all been crucial factors in the multiple uses of the past to shape contemporary politics, to justify violence and conflict, and to interpret the nature of intergroup relations. Precolonial nations are consolidated as "ethnicities" in modern Nigeria, reifying a link between the past and the present in ways that produce intense political debates and conflicts, including a three-year-long civil war. Nigerian politics is fierce, and those who control power dominate resources and often try to see their beliefs, even when crude and unarticulated, as the basis to move society forward. Daily political realities are chaotic,[31] and the presentation of histories reveals ambiguities, disharmony, and conflicts.

History becomes the tool of modern politics, as members of the political class turn to the past and to traditions to consolidate ethnic identities, and mobilize the people on the basis of communal histories, religions, and perceptions about others. Solid ethnic blocs define the boundaries of politics, creating a North–South cleavage. Although there are over two hundred ethnic groups in the country, only three have become politically dominant. The emergence of these three is tied to the use of history, the manipulation of traditions, the crude use of violence when necessary, and threats to secede. The minorities, too, have creatively turned histories of marginalization into the most powerful political discourse in contemporary Nigeria. Multiple histories and multiple identities reveal the absence of unity and harmony in the country. The contest for power is also about ethnic rivalries. Histories of discrimination instigate conflicts over resource allocation, minority rights, and power sharing. Histories of religious differences justify the extreme step of killing others, dehumanizing those on the other side. The civil war did not resolve the problems that led to it; there can be civil wars in the future.

History plays multiple roles: as past realities that shape contemporary politics; as a tool to be manipulated by political actors to attain desired results; as a tool to construct ideologies to institutionalize ideas about the "other" and to generate consensus within an "in group"; and as a weapon to counter the positions of others by way of presenting interpretations of events. History can be "used" and "misused" for political purposes, as we see in the successful case of turning to Oduduwa or the jihad to construct powerful political parties and ethnic constituencies.

The most partisan elements in Nigerian history derive from two interrelated elements: the use of Islam and the Caliphate to construct a power base in the North; and the historical reconstruction by southerners stressing the political dominance of the North—for example, describing the most powerful members of the northern political class as "The Kaduna Mafia." The so-called mafia is portrayed as ruthless, power greedy, and driven by opportunistic calculations to steal money. The history of the evolution of the "mafia" is tied to Islam and the dominance of the NPC in politics.[32] The country is presented as doomed if the power of the North is broken, terminated, and replaced by political managers drawn from the South.[33] In accusing the North, the chroniclers of history tend to ignore the fact that the South does the same, manipulating ethnicity, presenting alternative histories of past governments and war, all in order to justify claims to power and absolve themselves from the glaring failures of Nigeria.

The failure to manage and develop postcolonial Nigeria has created pressure to negotiate its survival as a nation. When civilians took over power in 1999 following a long period of military rule, the government set up the Oputa Panel to examine some past injustices by political leaders, cases of ethnic and religious violence, community rivalries, and matters relating to chaos and crises. The members of the public believed for a while that the Oputa Panel would work like the South African Truth and Reconciliation Commission. Although the panels sat and took evidence, it was hard to bring former military officers to testify and punish them. It was even harder to pay reparations to those who suffered. Even questions of punishment and restitution became difficult to resolve, and the Oputa Panel did not contribute to "sanitizing" a polluted political environment.

The national pastime in Nigeria is to discuss the origins of the country's problems; analysts blame the British, ethnicity, or the coup-loving generals. One ethnic group accuses another of bringing about corruption and decay. Regime change at the federal and state level is endemic, because of coups and countercoups, complaints about efficiency, and the failure of democracy. Each change turns the country into a "talking shop" with discussions on how to review the past in order to move forward, such that policies on taxation, revenue allocation, economic planning, and other matters of state are unstable.

In constructing a national history, academic and nonacademic scholars have advocated various views, some of which have legitimized state structures and institutions, and others that have advocated for ethnicity and secessions. There are scholars (and politicians) who think that "positive" historical narratives of the past could have overwhelmed the "negative" ones to create a united Nigeria. As if trying to reduce the credit to the British, some argue that Nigeria would have emerged as an evolutionary process of the natural coming together of its various "tribes."[34] The ties among the various precolonial groups—trade, religion, diplomatic contacts, and others, some argue, are sufficiently strong to sustain a united Nigeria. On the basis of this belief, universities mount courses on intergroup relations, as an academic justification for the power of history to unite and build nationalism. Ethnicity and pluralism, many scholars have argued, are not peculiar to Nigeria and should not be a hindrance to its survival and progress.[35] Many intellectual projects have sought to tap into history, and bring out its relevance to engineer the integration of the country. History, as a discipline, becomes invested with "wisdom" and offers "answers" to solve contemporary problems.[36]

Scholars, as well as politicians, have not been able to use knowledge and policies to solve the problems deriving from the history of Nigeria's creation, and the complications of managing a plural society. Nigeria is no longer under the military, but democracy is yet to solve the country's problems. On the contrary, democracy has reopened past wounds, bringing back memories of divisions. Communal and ethnic tensions are alive. The government is afraid to heed the call for a national conference, for fear of reopening old wounds and encouraging the call for secession and more states. To the government, a national conference will mean the coming together of nationalities who will use different histories to present alternative agendas on the future of Nigeria, threaten the survival of Nigeria, and create a platform for various groups to engage in a dogfight. To political leaders managing a fragile state, the past becomes dangerous, and the retelling of histories a source of problems for contemporary politics. The government that should be promoting history now sees the past as an enemy of the present.

Notes

1. Toyin Falola, *The History of Nigeria* (Westport, CT: Greenwood, 1999) provides an overview of the country's history. For a companion volume on the country's culture, see Toyin Falola, *Culture and Customs of Nigeria* (Westport, CT: Greenwood, 2000).
2. See, e.g., Kolawole Ogundowole, *Colonial Amalgam, Federalism and the National Question: A Philosophical Examination* (Lagos, Nigeria: Pumark, 1994).
3. F.A.O. Schwartz, *Nigeria: The Tribe, The Nation or the Race* (Cambridge, Boston: MIT Press, 1965).
4. S.E. Oyovbaire, *Federalism in Nigeria: A Study of the Development of the Nigerian State* (London: Macmillan, 1985).
5. Allison A. Ayida, *Rise and Fall of Nigeria* (Lagos, Nigeria: Malthouse Press, 1990).
6. Ismail A.B. Balogun, *The Life and Works of Uthman Dan Fodio: The Muslim Reformer of West Africa* (Lagos, Nigeria: Islamic Publications Bureau, 1975).
7. Abdulmalik Bappa Mahmud, *A Brief History of Shari'ah in the Defunct Northern Nigeria* (Jos: Jos University Press, 1988).
8. See, e.g., J. Tseayo, *Conflict and Incorporation in Nigeria: The Integration of the Tiv* (Zaria, Nigeria: Gaskiya, 1975).
9. C. S. Whitaker, Jr., *The Politics of Tradition: Continuity and Change in Northern Nigeria, 1946–66* (Princeton, NJ: Princeton University Press, 1970).
10. Ahmadu Bello, *My Life* (Cambridge: Cambridge University Press, 1962).
11. See, e.g., Mahdi Adamu, *The Hausa Factor in West African History* (Zaria, Nigeria: Ahmadu Bello University Press, 1978).
12. For an elaboration of the political strategy based on Islam, see Jonathan T. Reynolds, *The Time of Politics (Zamanin Siyasa): Islam and the Politics of Legitimacy in Northern Nigeria, 1950–1966* (Lanham, MD: University Press of America, 1999).

13. See, e.g., Rima Shawulu, *The Story of Gambo Sawaba* (Jos, Nigeria: Echo Publications, 1990).

14. Bello, *My Life*, 227.

15. See, e.g., Sheikh Abubakar Gumi (with Ismaila A. Tsiga), *Where I Stand* (Ibadan, Nigeria: Spectrum, 1992).

16. Matthew Hassan Kukah, *Religion, Politics and Power in Northern Nigeria* (Ibadan, Nigeria: Spectrum, 1993).

17. Toyin Falola, *Violence in Nigeria: The Crisis of Religious Politics and Secular Ideologies* (Rochester, NY: Rochester University Press, 1998).

18. See, e.g., Sam Babs Mala, ed., *Religion and National Unity* (Ibadan, Nigeria: Orita Publications, 1988).

19. Toyin Falola and Dare Oguntomisin, *The Military in Nineteenth Century Yoruba Politics* (Ile-Ife, Nigeria: University of Ife Press, 1984).

20. Toyin Falola and Olasope O. Oyelaran, eds., *Obafemi Awolowo: The End of an Era?* (Ile-Ife, Nigeria: University of Ife Press, 1988).

21. Toyin Falola, "Yoruba Town Histories," in Axel Harneit-Sievers, ed., *A Place in the World: New Local Historiographies from Africa and South Asia* (Leiden: Brill, 2002), 65–86.

22. Adiele Afigbo, *Ropes of Sand: Studies in Igbo History and Culture* (Ibadan, Nigeria: University Press Ltd., 1981).

23. Nnamdi Azikiwe, *My Odyssey: An Autobiography* (London: C.Hurst, 1970).

24. Paul Obi-Ani, "Post-Civil War Nigeria: Reconciliation Or Vendetta?," in Toyin Falola, ed., *Nigeria in the Twentieth Century* (Durham: Carolina Academic Press, 2002), 473–483.

25. Nsukka Analyst, *Marginalisation in Nigerian Polity: A Diagnosis of "The Igbo Problem" and the National Question* (Enugu: Hillys Press, 1994); Pat Uche Okpoko, "Three Decades After Biafra: A Critique of the Reconciliation Policy," in Falola, ed., *Nigeria in the Twentieth Century*, 486–496.

26. See e.g., Isidore Okpewho, *Once Upon a Kingdom: Myth, Hegemony, and Identity* (Bloomington, IN: Indiana University Press, 1998).

27. *Nigeria, Report of the Panel Appointed by the Federal Military Government to Investigate the Issue of the Creation of More States and Boundary Adjustments in Nigeria* (Lagos: Federal Government Printer, 1975), 13.

28. R.T. Akinyele, "Strategies of Minority Agitations in Nigeria, 1900–1954," *Nigerian Journal of Inter-Group Relations*, 1: 1 (1995), 1–9.

29. Matthew H. Kukah, "Minorities, Federalism and the Inevitability of Instability," a paper presented at the Legal Research and Resource Development Center, Lagos, February 14–15, 1994.

30. Anthony Agbali, "Politics, Rhetoric and Ritual of the Ogoni Movement," in Falola, ed., *Nigeria in the Twentieth Century*, 507–531; Claude E. Welch, "The Ogoni and Self-Determination: Increasing Violence in Nigeria," *The Journal of Modern African Studies*, 1: 33 (1995), 635–649; Ken Wiwa, *In the Shadow of a Saint: A Son's Journey to Understand his Father's Legacy* (South Royalton, VT: Steerforth Press, 2001); Craig W. McLuckie and Aubrey McPhail, *Ken Saro-Wiwa: Writer and Political Activist* (Boulder, CO: Lynne Rienner, 2000); Onookome Okome, ed., *Before I am Hanged: Ken Saro-Wiwa, Literature, Politics and Dissent* (Trenton, NJ: Africa World Press, 2000); and Abdul Rasheed Na'Allah, ed., *Ogoni's Agonies: Ken Saro-Wiwa and the Crisis in Nigeria* (Trenton, NJ: Africa World Press, 1998).

31. Karl Maier, *This House Has Fallen: Midnight in Nigeria* (New York: Public Affairs, 2000).

32. Bala J. Takaya and Sonni Gwanle Tyoden, eds., *The Kaduna Mafia: A Study of the Rise, Development and Consolidation of a Nigerian Power Elite* (Jos, Nigeria: University of Jos Press, 1987).

33. Okechukwu Okeke, *Hausa–Fulani Hegemony: The Dominance of the Muslim North in Contemporary Nigerian Politics* (Enugu, Nigeria: Acena Publishers, 1992).
34. See, e.g., B.J. Dudley, *An Introduction to Nigerian Government and Politics* (London: Macmillan, 1982).
35. Onigu Otite, *Ethnic Pluralism and Ethnicity in Nigeria* (Ibadan, Nigeria: Shaneson, 1990).
36. Ukwu I. Ukwu, ed., *Federal Character and National Integration in Nigeria* (Kuru, Nigeria: National Institute for Policy and Strategic Studies, 1987).

Suggestions for Further Reading

Falola, Toyin, *The History of Nigeria* (Westport, CT.: Greenwood, 1999).

Falola, Toyin, ed., *Nigeria in the Twentieth Century* (Durham: Carolina Academic Press, 2002).

Falola, Toyin, *Violence in Nigeria: The Crisis of Religious Politics and Secular Ideologies* (Rochester: University of Rochester Press, 1998).

Forrest, Tom, *Politics and Economic Development in Nigeria* (Boulder, CO: Westview Press, 1993).

Maier, Karl, *This House Has Fallen: Midnight in Nigeria* (New York: Public Affairs, 2000).

Osaghae, Eghosa E., *Crippled Giant: Nigeria Since Independence* (London: Hurst and Company, 1998).

PART 3

New Lessons

CHAPTER 9

Histories and "Lessons" of the Vietnam War

Patrick Hagopian

In what Maureen Dowd remembers as the "innocent summer" before September 11, 2001, the U.S. defense secretary Donald Rumsfeld sponsored a study of ancient empires: Macedonia, Rome, and the Mongols.[1] At a time when intellectuals had begun to speak unashamedly about an American empire, Rumsfeld evidently sought lessons from Alexander the Great, Julius Caesar, and Genghis Khan on how previous empires retained their dominance. In the aftermath of the attacks on the Pentagon and the World Trade Center, Rumsfeld's Pentagon began organizing screenings of Gillo Pontecorvo's *The Battle of Algiers*, which showed French troops battling Arab nationalists organized in secret cells—just as Nixon White House staff had watched the same film 30 years earlier to get pointers on combating urban terrorism.[2] These attempts to learn the "lessons" of history may appear naïve to some, sinister to others, but they were not unique or even unusual. In the United States, since World War II, politicians have often used historical examples to arrive at, or to explain, foreign policy decisions.

In selecting historical precedents, U.S. policymakers have tended to refer to unfortunate events as touchstones for what must never be repeated. In the second half of the twentieth century, the ruling precedents were two contending historical warnings: the 1938 Munich agreement, which granted Adolf Hitler a third of Czechoslovakia in the mistaken belief that he would be appeased, and the Vietnam War. For the World War II generation that led the United States in the early years of the cold war, the example to be avoided was "another Munich." Policymakers of this generation believed that aggression must be opposed steadfastly and early, and that yielding ground did not satisfy dictators' ambitions but rather fueled them. For the generation that made policy after 1975, the Vietnam War offered "lessons" that contrasted with those drawn from Munich. The "Vietnam syndrome" cautioned against becoming drawn into conflicts abroad where American vital interests were not immediately at stake. The "lessons" of Vietnam and Munich thus pointed in very different, indeed in opposite, directions. The "lesson" of Munich was that one should not wait until one's vital centers were threatened before being compelled to act, whereas the "lesson" of Vietnam was that one should not act hastily unless a vital interest was at stake.

Although the "lessons" of Vietnam appeared to prevail during the 1980s, they were highly contested. Whereas almost everyone agreed that "another Vietnam" should be avoided, there was a great deal of disagreement about what that phrase meant. For some politicians, "another Vietnam" meant an unnecessary intervention abroad; for others, it meant defeat resulting from a loss of will. In the 1980s, Ronald Reagan countered the "Vietnam syndrome" by attempting to redefine the Vietnam War or by denying its pertinence to the contemporary context of Central America. Reagan also applied other historical precedents, notably the Munich analogy. As we shall see, the Munich analogy never went out of favor entirely, and it has lately experienced a new vogue.

The Munich Analogy

For America's post-1945 cold warriors, Prime Minister Neville Chamberlain's failure at Munich to resist the German takeover of the Sudetenland demonstrated that one should not wait until an enemy was at one's doorstep before countering his advances. In 1938 Chamberlain declared to a grateful Britain that he had achieved "peace in our time." Germany was not satiated by its success at Munich, though; indeed, it was emboldened, and a general war soon followed.

For American policymakers who applied the "lesson" of Munich in the post–World War II period, the Soviet Union stood in for Germany as a "totalitarian" power. American strategists determined that one could not appease dictators and that it was better to stop their aggression early rather than late. When war broke out in the Korean peninsula, Harry Truman remembered the weak responses of the democracies in the 1930s to aggression by Germany, Italy, and Japan. As Truman saw it, the lesson of the 1930s was that fighting in Korea was necessary in order to avoid World War III.[3]

President Dwight D. Eisenhower referred to the lessons of Munich in attempting to persuade Britain that Vietnam must not fall to the communists.[4] He worried that surrendering the offshore islands Quemoy and Matsu to China would be a "Western Pacific Munich."[5] When Lyndon Johnson met Eisenhower in February 1965 to solicit his views about Vietnam, Eisenhower used the Munich analogy to explain that negotiations would not lead to a favorable outcome and that force of arms must prevail instead.[6]

When Lyndon Johnson's administration expanded U.S. military involvement in Vietnam, his advisers drew conflicting "lessons" from Korea and Munich. The Joint Chiefs of Staff pointed to the "lesson" of the Korean War, which had eroded public support for Truman's presidency and ended in stalemate. The lesson of Korea, they said, was that the United States should never again fight a land war in Asia.[7] But Johnson drew a different lesson: stalemate in Korea had cost Truman and his secretary of state, Dean Acheson, their effectiveness. The "lesson" of Korea was that the United States must be firm and "hold the line" in Vietnam.[8] The "lesson" of Munich reinforced this conclusion, and diplomats and military advisers invoked the Munich precedent in explaining why it was necessary to halt communist advances in Southeast Asia. As President Johnson was making crucial decisions about expanding America's military effort in 1964 and 1965, ambassador to South Vietnam Henry Cabot Lodge and Air Force Chief of Staff Curtis LeMay advised him that Vietnam was a test of America's will to resist communism and that he must not fail the test in a repeat of the Munich agreement.[9] At a 1965 press conference, Johnson declared: "This is the clearest lesson of our time. From Munich until today we have learned that to yield to aggression brings only greater threats."[10] Johnson repeated the point in another address to the nation: "we learned from Hitler at Munich that success only feeds the appetite of aggression."[11] In the spring of 1965, when students organized the first teach-ins in opposition to the Vietnam War, the Johnson administration released a propaganda film titled *Why Vietnam?* to explain why the government was sending young Americans to risk their lives in a remote and distant part of the world. The film gave the Munich analogy pride of place in its argument.

Near the start of the film, Lyndon Johnson repeatedly asks, "Why Vietnam?" and we see Hitler riding in an open-topped car, greeted by cheering crowds. The voice-over responds to the question with the words, "Munich 1938." Cut to Chamberlain disembarking from an aircraft, greeted by an honor guard of German soldiers, the voice-over continuing, "This meeting will long be remembered, for it opens the door to the dreams of dictatorship." After Chamberlain reads his announcement of the agreement with Hitler, the voice-over says, "Peace in our time: a shortcut to disaster." We see shots of devastated cities and the voice-over refers to the democracies' weakness at Munich, and to their failure to halt other aggressions by Hitler and the Italian dictator Benito Mussolini. When the camera returns to Lyndon Johnson, he explains the "painful lessons" that retreat does not bring safety and weakness does not bring peace.[12]

A decade later, as the Vietnam War was drawing to a close, American leaders continued to invoke the lessons of Munich. In October 1974, President Ford said, "As a young man in the 1930s, I remember the isolationism that blinded so many Americans to the menace of Hitler's Germany and its totalitarianism." Americans of his generation, Ford said, vowed never to repeat the same mistakes and to keep America strong—that was the basis on which the United States had constructed an international network of mutual security. However, the assumptions that had governed American policy were now being challenged. Ford reported that in the past decade, the "whole fundamental policy of mutual security and strength" had come under attack.[13]

If the logic of the Munich analogy helped impel the United States to war in Indo-China, the Vietnam War's duration, its unpopularity, and its unsuccessful outcome helped to discredit Munich as a touchstone of policy. By the time the United States withdrew its last combat forces from Vietnam, the world in which President Johnson had made his decisions for war had been transformed. Henry Kissinger and Richard Nixon's "opening to China" made balance-of-power politics possible and altered the premises of cold war confrontation, enabling détente with the Soviet Union. Kissinger saw a scaling back of U.S. international commitments as the only viable alternative to a new, war-weary isolationism.[14] As national security adviser and secretary of state, he wanted a freedom of maneuver that was foreclosed by the obligatory confrontational politics bound up with the Munich analogy.

Liberal internationalists such as Senator William Fulbright also challenged the cold war orthodoxies inherent in the "lessons" of Munich. Fulbright questioned what he called the United States' "arrogance of power" and argued that the United States should scale down its strategic goals and responsibilities. The United States should not presume that it had the responsibility to crusade around the world for values such as "freedom" and "democracy." Instead, it should pursue a more circumscribed set of strategic interests.[15]

America's leaders quickly tried to consign the Vietnam War to oblivion. Asked soon after the collapse of South Vietnam and the evacuation of the American embassy in Saigon to explain what had gone wrong, Secretary of State Henry Kissinger said that it was not the right time to make an assessment. What the country needed now, he argued, was "to put Viet-Nam behind us and concentrate on the problems of the future."[16] President Ford declared on April 29, 1975, that the evacuation of Americans and Vietnamese from South Vietnam to the fleet in the South China Sea "closes a chapter in the American experience."[17] Ten days later, he said: "The war in Vietnam is over . . . And we should focus on the future. As far as I'm concerned, that is where we will concentrate."[18]

If the outcome and the aftermath of the Vietnam War had been otherwise, the Munich analogy might have been vindicated. After all, the United States had made a commitment to South Vietnam and had demonstrated that it was willing to assume enormous costs and impose even greater costs, in human lives, on its enemies. But the Vietnam War divided the United States domestically; the nation also suffered international setbacks in the 1970s with the emergence of Soviet-backed governments in southern Africa, Ethiopia, and Central America, and the loss of a key ally in Iran. Observers blamed the United States' weakness on the "Vietnam syndrome," arguing that the war-weariness of the American public made the United States less able to defend its interests abroad. The recognition that overcommitment by the United States might be just as harmful to its interests as being overly timid undermined the "lesson" of Munich.

Post–Vietnam War caution was in keeping with the logic of the "Nixon Doctrine," which asserted that the United States would aid and equip its allies, but that they would have to do the fighting.[19] Before the fall of Saigon, Henry Kissinger had recommended, and the United States had pursued, covert programs to undermine the Marxist MPLA movement in Angola and to destabilize Salvador Allende's government in Chile.[20] The pattern was set: the United States would be more cautious in committing its troops to military actions that would result in American dead and that would therefore become domestically unpopular. In what Nixon had called a "revised policy of involvement," the United States would continue to confront its enemies, but indirectly, and the bodies the conflict generated would be foreign, not American.[21] This parsimony in the shedding of American blood matched the public's preferences. In January 1975, when Kissinger was so bold as to declare that there were some causes (e.g., oil) for which America might employ its own forces, editorial writers around the country chided him with his own rendition of the lessons of Vietnam. The *Washington Post*, for example, repeated that the "great lesson" of that war was that it was easier to get into wars than to get out of them.[22]

"No More Vietnams"

From the moment the South Vietnamese government forces collapsed in 1975, different "lessons" about the Vietnam War contended with one another. Most Americans agreed that the United States should avoid another Vietnam War, but they disagreed about what had gone wrong in Vietnam. As German politicians did after World War I, Gerald Ford promulgated a "stab-in-the-back" myth, blaming Congress for the defeat. On April 3, 1975, Ford called upon future historians to decide where the blame for the impending defeat of South Vietnam should lie, giving them a hint by pointing to Congress's refusal to provide South Vietnam with military supplies. But this was a cynical effort to hand Congress responsibility for a defeat that occurred during Ford's watch.[23]

In subsequent years, Henry Kissinger, ambassador to South Vietnam Graham Martin, and others also blamed Congress for the catastrophic collapse of South Vietnam.[24] Blaming Congress would preserve the illusion, fostered by President Nixon, that the United States had achieved "peace with honor" through the 1973 Paris Peace Accords. "Congress refused to fulfill our obligations," Nixon wrote, and its "tragic and irresponsible action" was to blame for the defeat of the South Vietnamese government by an invasion from North Vietnam in April 1975.[25] Demonstrating that there was plenty of blame to go around, Nixon and Kissinger blamed other members of the administration, public opinion, the press, intellectuals, and the antiwar movement for preventing them from using sufficient force to win a military victory.[26]

The litany of "what might have beens" pointed the finger at culprits other than the U.S. Congress, especially the Johnson administration. General William Westmoreland argued in 1976 that the U.S. military was not to blame for the defeat because it "quite clearly did the job that the nation asked and expected of it." Hanson Baldwin wrote that it was

> clear that the blame for the lost war rests, not upon the men in uniform, but upon the civilian policy makers in Washington—those who evolved and developed the policies of gradualism, flexible response, off-again-on-again bombing, negotiated victory, and, ultimately, one-arm-behind-the-back restraint and scuttle-and-run.[27]

Colonel Harry Summers repeatedly told a story of his exchange with a North Vietnamese colonel after the war, when the colonel agreed that North Vietnam "never defeated [the USA] on the battlefield."[28] The lesson was clear: the defeat rested on the shoulders of the politicians, especially the gradualist policies of the Johnson administration.

The majority of the public was, however, unconvinced by the efforts of politicians and military officers to exonerate themselves by blaming others. The lessons that the public learned from the Vietnam War were that politicians and military commanders could not be trusted to tell the truth in wartime and that their optimistic predictions were often false. The public also learned that it was easier to get into a war than to get out of one, once it starts to go badly. An initial commitment of U.S. troops inevitably wagered America's prestige and "credibility" on the outcome of a struggle. If the U.S. forces could not achieve a quick victory or tangible gains, more troops might be required to shore up a weak position; each fresh deployment increased the invest-ment of America's credibility. Thus, the public concluded that the nation should be extremely cautious about embarking on any commitment of troops abroad, however small, because it could suck the nation into "another Vietnam," in which the eventual alternatives were humiliating defeat or endless, unsustainable bloodshed. "No more Vietnams" became a rallying cry for opponents of American interventionism.

During his presidential campaigns of 1976 and 1980, Ronald Reagan undertook a crusade for renewal. As Richard Melanson has written, the neoconservatives who promoted Reagan's anticommunist foreign policy

> correctly realized that the legacy of Vietnam required revision before a revitalized policy of anti-Communist containment could be adopted. So long as substantial numbers of Americans continued to believe that interventions inevitably produced Vietnams, it would be difficult to significantly reassert American power. Only by convincingly altering the "lessons of Vietnam" could a new Administration hope to gain domestic support for global containment.[29]

Debating the "Lessons" of Vietnam

When Ronald Reagan campaigned for office, the question he confronted was, how can any politician alter the nation's memory? The connected themes of conservative revisionism were that the war had been worthwhile; that it could and should have been won; and that the reasons for defeat were poor strategy and a loss of will. In the conservative view, intervention in Vietnam had supported a vital U.S. interest. The problem with the war was that it was prosecuted ineffectively, not that its purposes were misguided or its conduct immoral. America's goals should have been more clearly defined, the war should have been fought more vigorously, and the public mobilized behind the war effort.[30] What had gone wrong in Vietnam, the conservatives argued, was not that the United States fought in Vietnam but rather that it had abandoned its ally. "Another Vietnam," in their interpretation, would be another failure to stay the course. "No more Vietnams" meant not that the nation should be cautious about committing its forces but that it should renew its faith in itself and redouble its resolution in facing down its communist antagonist.[31]

In August 1980, Ronald Reagan told an audience of military veterans that the Vietnam War had been a "noble cause" that "should have been won."[32] Five years after the Vietnam War ended with the defeat of America's South Vietnamese ally and the unification of Vietnam under a communist government, he argued that the "Vietnam syndrome" had made American foreign policy timid, allowing the Soviet Union to expand its influence around the world. In another echo of the "stab-in-the-back" myth, he suggested that the U.S. government's loss of will and determination had betrayed the war effort: "Let us tell those who fought in that war that we will never again ask our young men to fight and possibly die in a war our government is afraid to win."[33] The audience broke into "sustained and boisterous cheers."[34] The solution Reagan proposed in his speeches was for Americans to reassert their patriotism and pride, and for the United States to strengthen its military forces.

In office, Reagan continued to insist that the government must never prevent U.S. forces from winning their wars.[35] Other Republican politicians echoed this theme. At the dedication of the Vietnam Veterans Memorial in November 1982, Senator John Warner said "we should never again ask our men and women to serve in a war that we do not intend to win." Among the other "sober lessons of history" learned in Vietnam that Americans must "never forget" were these:

> We learned that we should not enter a war unless it is necessary for our national survival. We learned that, if we do enter such a war, we must support our men and women to the fullest extent possible.[36]

Reagan's defense secretary, Caspar Weinberger, made the same points during a Veterans Day ceremony at Arlington National Cemetery. In almost identical terms, he said that the "terrible lesson" of the Vietnam War was that "we should never again ask our men and women to serve in a war that we do not intend to win."[37]

Reagan determined not simply to halt but to reverse the advances by Soviet-allied forces in the Third World. In order to do so, though, his administration had to overcome public and Congressional fears of "another Vietnam," as well as legislators' reluctance to cede the foreign policy prerogatives that they had clawed back in the waning days of the Nixon administration by passing the War Powers Act, legislation intended to reinforce the war-making authority that the Constitution assigned to Congress. The problem for the Reagan administration was that there was no clear domestic consensus in favor of a more active foreign policy. Any military involvement in regions where American interests encountered armed opponents evoked the "Vietnam trauma."[38]

Just as the "lessons of the past" influenced present policymaking, new events could help to reinforce or counter prevailing interpretations of the past. Secretary of state Alexander Haig was aware of this when, in assessing the consequences of the Vietnam War, he said that America's loss of confidence and faltering leadership appeared to make the nation unable to defend its international interests. Haig's solution: confidence in the United States "must be reestablished through a steady accumulation of prudent and successful actions."[39] Such confidence-building successes would gradually inure the American public and politicians to the debilitating Vietnam syndrome, in a sort of behavioral therapy aimed at boosting U.S. morale as much as achieving a particular local objective.

The October 1983 U.S. invasion of Grenada seemed to fit the psychological need that Haig identified, as a military operation that achieved its goals with few U.S. lives lost. The invasion of the Caribbean island was just the sort of event that would hearten the American public to the fact that military operations could bring victory, not just shame, grief, and defeat. But events in Lebanon that same October demonstrated the risks that military operations entailed: a suicide bomber attacked the headquarters of a Marine detachment; 241 Marines died. Soon after, Reagan withdrew the Marines—or, as he delicately put it, "redeployed" them to U.S. ships offshore. The euphemism fooled no one; the hasty redeployment cost Reagan dearly in his public approval ratings.[40] Yet Reagan's inconsistent actions and behavior—his bellicose talk, his wish for swift victories, and his alarm at setbacks—seemed perfectly to match the public's contradictory impulses, its hunger for a renewed pride in the nation, but its persistent risk-aversion.[41]

The following year, referring to Grenada, Reagan declared in his reelection campaign that America was "standing tall" again. The Lebanon debacle did not seem to undercut his electoral support and he won by a landslide; nevertheless, the Beirut bombing reinforced the "Vietnam Syndrome" and made the public still more wary about the risks of sending in the Marines to any international trouble spot. Reagan's ideological commitment to a strong military and to militant anticommunism meant that, while he was personally well-liked by much of the public, he was never entirely trusted. Much of the public feared he would lead the nation into another Vietnam-like "quagmire," and the sincerity of his anticommunist beliefs increased these fears. Reagan's determination to revive the cold war set off an ideological debate that awakened the impassioned arguments of the Vietnam War era. Henry Kissinger, Gerald Ford, and Jimmy Carter had all spoken of the nation's need to put the con- flicts of the Vietnam era behind it and to turn over a new page. By trumpeting the rightness of the war and gainsaying those who doubted its morality, Reagan attempted to end debates and uncertainties, not by downplaying differences in the name of healing but by asserting the rightness of one view of the war over another. But Americans had come by their views of the Vietnam War over long years of fighting that had been surrounded by public debate. Forged in the reflected heat of warfare, and tempered by blood and tears, their beliefs about the Vietnam War melded strong emotions and ideas. Political speeches, no matter how rhetorically effective, had little impact on such powerfully wrought attitudes. Stirring the embers of wartime arguments by declaring the war to have been a "noble cause" did not change people's minds but merely intensified their existing views.

In response to Reagan's "noble cause" speech, some newspaper editorial writers applauded him but others were aghast. The *Denver Post* editorialized, "Reagan is right. As individuals and a nation, we generally fought for noble motives in Vietnam."[42] The *Arizona Republic* said "Reagan deserves high marks" for calling the war a "noble cause."[43] For every editorial that applauded his courage and frankness, though, another condemned Reagan for sounding naïve. The "noble cause" idea, according to the *Miami Herald*, might have been attractive to some, but was "regret- tably divisive, simplistic, and wrong."[44] The *Chicago Sun-Times* elaborated, cataloguing a train of U.S. deceptions and crimes in Vietnam:

> Who can forget official U.S. duplicity in the Diem coup, the "secret wars" in Cambodia and Laos, the falsified bombing records, the fragmentation weapons against civilian targets?
>
> And the defoliants? "Hearts and minds?" [The] My Lai [massacre]? The lies to the American Congress and the American people about the scope of the war and the prospects for peace? Noble?[45]

The political problem that the Reagan administration faced in the 1980s was an unusually complicated instance of the classic hegemonic task of winning the consent and overcoming the resistance of society's component groups in constructing a ruling bloc. Described by Gramsci as the operation of power in the creation of a series of shifting equilibria, this balancing act, the achievement of hegemony, was made all the more difficult by the fissured ground on which it was performed.[46] Reagan sought in the "noble cause" speech to rally his core, pro-military constituency; but Reagan's handlers were always keen to ensure that he avoid alienating a significant part of the electorate except when taking positions on matters that they and the candidate thought essential. Reagan's vindication of the Vietnam War appeared, on these grounds, to backfire. Although it galvanized those who had supported the war effort, it provoked those who had opposed it, polarizing Americans along the lines of their war-era divisions.

Talk of "another Vietnam" permeated the debates about Reagan's foreign policy. Reagan's foreign policy advisers were determined to hold the line against communist advances in the "Third World," and to begin to roll them back. Reagan's Democratic opponents in Congress, though, feared that the administration's eagerness to overcome the "Vietnam syndrome" would lead it to engage U.S. forces in an ill-advised venture. The most likely arena for U.S. involvement to increase gradually until the nation was involved in a Vietnam-like quagmire was Central America, where U.S. military "advisers" were supporting the Salvadoran government—just as the first American forces in South Vietnam had been "advisers."

The Reagan administration's inner circle was divided about raising the rhetorical and political stakes in El Salvador so as to justify increasing military aid to the region. Reagan's aides were aware of the political costs this involvement could exact, by provoking the hostility of his Congressional opponents and by diminishing the public's support of his presidency. These fears were justified. Public opinion polls revealed, and the White House's privately commissioned polls confirmed, that during his first year in office Reagan's Central America policy did undermine his popularity.[47] The bad news continued in his second year. In early 1982, polls showed public opposition to Reagan's policies in Central America rising sharply, with almost two-thirds of respondents to a Harris poll disapproving Reagan's handling of El Salvador.[48] In February 1982, a Gallup poll found that 74 percent of a sample of the public who knew which side the United States was backing said that it was likely that U.S. involvement in El Salvador could turn into "a situation like Vietnam—that is, that the United States would become more and more involved as time goes on."[49]

A majority of the public also disagreed with the Reagan administration's policy toward Nicaragua, where the administration was supporting the anticommunist *contras* (counterrevolutionaries).[50] Congress began to assert its policymaking prerogatives, and in its decisions about appropriations of funds for foreign policy purposes, placed limits on administration aid to the Nicaraguan *contras*.[51] In December 1982, the House of Representatives passed the Boland Amendment, banning covert aid to assist in another of Reagan's goals in his Central America crusade: the overthrow of the socialist Sandinista government in Nicaragua.[52] In the debate leading to the adoption of the amendment, Representative George Miller (Democrat of California) said, "Some of us came here to stop [the] Vietnam [War]. . . And here is a chance to stop the new one."[53] Meanwhile, the leftist rebels in El Salvador launched offensives in the autumn of 1982 and renewed them in January 1983, encouraged by political turmoil in the Salvadoran government and security forces. An administration official pointed to the fragility of the Salvadoran government and warned that any sign of a weakening commitment to them on the part of the U.S. government might set off a panic: "If the sense spreads that the U.S. will desert them, I don't know what they'll do. It's Vietnam all over again."[54]

The shadow of Vietnam continued to hang over Reagan's Central American policy in the remainder of his first term in office. Nine Gallup and Harris polls spanning

Reagan's first administration indicated that majorities ranging from 62 to 77 percent agreed that the conflicts in Nicaragua and El Salvador might involve the United States in "another Vietnam." The Gallup poll questions glossed "another Vietnam" by describing it as a place where "the U.S. would become more and more deeply involved as time goes on."[55]

Facing public and Congressional opposition, the President addressed a joint session of Congress in April 1983, attempting to convince legislators that the Nicaraguan government threatened to export totalitarianism to its neighbors.[56] But he was unable to win consistent Congressional support for his policies. The problem for Reagan was that the more he asserted the importance of Central America as an area where the United States had vital national interests, the more listeners feared he might take the nation to war there. The more he tried to assuage fears of "another Vietnam" by trying to redefine Vietnam as a noble cause worth fighting, the more listeners worried that he was willing to launch a repetition. Reagan tried to calm such fears by insisting that there was no parallel whatever between Central America and Vietnam, but his own presentation of a world-historical struggle between the forces of freedom and of totalitarianism undermined this claim.[57] In March 1984, Senate liberals opposed to the introduction of U.S. troops in Central America made passionate appeals to their colleagues not to allow another Vietnam. Reagan criticized the Democrats for failing to support his policies in Central America, saying in May 1984, "Unfortunately, many in Congress seem to believe that they are still in the troubled Vietnam era, with their only task to be vocal critics, not responsible partners."[58] In October 1984, Congress again suspended *contra* aid by passing the second Boland amendment, with frequent references in the debates to the danger that U.S. involvement in Central America might lead to "another Vietnam."

Reagan's advisers, particularly Michael Deaver, were extraordinarily sensitive to the tides of public opinion and recognized that foreign policy was a politically dangerous area for their boss. Rhetoric about the "noble cause" had earned Reagan the reputation as trigger-happy extremist. The White House's privately commissioned surveys found that Reagan's speech in which he referred to the Soviet Union as an "evil empire," a statement countenancing limited nuclear war, and an off-the-cuff remark about beginning bombing the Soviet Union in five minutes had all "put some Americans' teeth on edge."[59] Reagan's "pragmatic" advisers—principally Deaver and James Baker—were just as worried as the public when Reagan sounded off about Central America. They tried to rein in Reagan's anticommunist predilections, and feared that "crazies" in the administration would circumvent Congressional restrictions on funding the *contras* (fears that were well-founded, as the furor about White House aide Oliver North's illegal operation to funnel money to Central America, the Iran-Contra scandal, eventually revealed).[60]

The public's wish to avoid "another Vietnam" was not only founded on fears of a long, drawn-out conflict, but was reinforced by questions about the war's morality. A series of Gallup polls taken between 1978 and 1986 showed that some two-thirds to three-quarters of the poll samples felt that the Vietnam War was "more than a mistake," but was "wrong" or "fundamentally wrong and immoral."[61] These responses demonstrate how hard a task was Reagan's effort to redefine the war as noble. Fortuitously, a poll that asked a leading question with a variant wording allows us to measure the degree of public resistance to Reagan's statement that the Vietnam War was a "noble cause." In 1985, a pair of polls taken a few months apart cued respondents to consider alternative judgments: some people, the preamble to the question stated, thought that the war was a "noble cause"; others said that it was wrong and immoral. With which assessment did the respondents agree? Despite the prompt in the wording of the question, only one-third of the sample agreed with the idea that the war was a noble cause. Although the number who felt the war was wrong, or wrong and immoral, diminished in comparison with other polls, it still constituted a plurality of the responses.[62] As these polls demonstrate, a substantial number were too strongly convinced of the war's wrongness to be swayed by Reagan's alternative judgment. Consequently, Reagan began to back off from strident declarations about the rightness of the Vietnam War until near the end of his term of office, when he declared that the cause for which the troops had fought in Vietnam was "just."[63]

The political parties' auxiliary forces skirmished on the terrain of Vietnam's "lessons." Policy debates were conducted, more or less obliquely or in parallel, through arguments about what the proper "lessons" of Vietnam were. The parallel was clearest in Richard Nixon's *No More Vietnams*. He argued that "another Vietnam" should not be defined as another military intervention in the Third World; nor should it be defined as an immoral war, since the Vietnam War was "morally right." Instead, "another Vietnam" meant simply, military defeat: Nixon wrote, " 'No More Vietnams' can mean that we will not *try* again. It *should* mean that we will not *fail* again [emphases in original]."[64] For the Democrats, though, the lessons of Vietnam recommended prudence in the use of force. Stephen Solarz, a Democratic Congressman, said that the Vietnam experience was "salutary in the sense that it made us much more cautious and careful in situations where vital U.S. interests were not involved."[65] There were differences of opinion even among Republicans, with some more willing to adapt to the cautionary lessons of Vietnam and others more prone to repudiate them. John Negroponte, a diplomat with service in several Republican administrations, sounded a similar note to Solarz when he said that the Vietnam War demonstrated that "even for a country so wealthy and powerful as the United States, there were costs beyond which it would be extremely difficult for us to go. I think it has had a moderating effect on our behavior, if you will, a tempering effect."[66]

The Powell Doctrine

In 1984, defense secretary Caspar Weinberger blended the competing "lessons" of the Vietnam War—that the United States should be strong and determined in combating its enemies, but cautious about where and when to do so—in a series of tests governing the use of military power. Later, Colin Powell, as Chairman of the Joint Chiefs of Staff and as Secretary of Defense, appropriated them wholesale, so that Weinberger's principles have come to be known recently as the "Powell Doctrine."[67] According to the doctrine, the U.S. government should not send forces into combat except in pursuit of a vital interest, and then only as a last resort; if it employed force, it should do so wholeheartedly, with the clear intention of winning; it should define its objectives clearly, and should send sufficient forces to accomplish them; it should commit combat forces only when there was "reasonable assurance we will have the support of the American people and their elected representatives in Congress," and it should communicate candidly with them about the potential costs and risks involved.[68]

The principle that force should be used overpoweringly, if at all, may give Weinberger's principles the superficial cast of a fiercely militaristic document. In fact, Weinberger's was an extremely cautious statement, imposing strict limits on the circumstances in which the United States would use its armed forces. It is rarely possible for a military power, even one as preponderant as the United States, to guarantee the swift and successful outcome of a conflict; and even if an initial consensus for military action exists, there is no guarantee that it will persist over time in the face of setbacks, moral qualms, or slow progress. As such, the Weinberger principles did not supersede the Vietnam syndrome but rather codified it so that policymaking could be conducted on a more predictable and rational footing. In return for a hoped-for consensus, Weinberger announced restrictions on the use of force that no predecessor had observed. As Max Boot points out, the United States has frequently committed troops to combat without a vital national interest, without significant popular support, without declarations of war, without exit strategies, and in settings that require U.S. troops to engage in nation-building.[69]

Each tenet of the Weinberger principles codified one of the negative "lessons" of Vietnam. But this response to the "lessons" of Vietnam did not emerge in a political vacuum—it arose in the context of the 1980s and addressed the problems the Reagan administration was experiencing in gaining support for its foreign policy. The fear that the president would maneuver the nation into war without a national consensus in favor of it made the failure of consensus in Vietnam relevant to the 1980s and required Weinberger's enunciation of a corrective principle.

Meanwhile, alongside the doctrine of "overwhelming force," U.S. military planners developed other approaches to the use of force in international affairs. In the 1980s, the U.S. post–Vietnam War reluctance to commit its own troops abroad was codified under the heading of "Low Intensity Conflict," an umbrella term that covered covert actions, counterinsurgency operations, and the support for pro-American rebellions, insurgencies, and proxy wars.[70] This extended the Nixon Doctrine's principle that in the long, twilight struggle against the United States' adversaries, it was better that Americans should not shed their blood, but instead should arm and equip others to do so for them. As director of Central Intelligence William Casey remarked, "It is much easier and less expensive to support an insurgency than it is for us and our friends to resist one. It takes relatively few people and little support to disrupt the internal peace and economic stability of a small country."[71] The Reagan administration put that principle into effect through its support of the *contras'* insurgency in Nicaragua and the anti-Soviet forces in Afghanistan and Angola.

The main tenets of the Weinberger principles and the Powell Doctrine—that overwhelming force must be employed directed toward finite and clearly defined goals backed by a public and Congressional consensus—have united most Republicans and Democrats since the mid-1980s. There was little disagreement on these points during the Congressional debate about the 1991 Gulf War, and most American editorial writers and commentators in the debates about the use of military force there and about military intervention in former Yugoslavia broadly accepted them. This emerging consensus about the *conditions* governing use of force does not imply, though, that there is now broad agreement about the *desirability* of its use or the circumstances under which that would be appropriate.

Since the enunciation of the Weinberger principles, the United States has usually been hesitant in its commitment of ground forces except where the odds of success weighed heavily in its favor. However, the Clinton administration did not accept the principles as universally valid. Under President Clinton, there was no longer a single, overarching concept that would govern America's use of force beyond its borders. In Haiti and Bosnia, for example, the Clinton administration sent troops into high-risk and combat situations without a clear public consensus in favor of those deployments and under circumstances where the missions could well have turned sour. Clinton's ad hoc approach to foreign policymaking paralleled a lack of consensus in the 1990s about which analogy, Munich or Vietnam, would prevail. Throughout that decade, pundits invoked both precedents, debating whether either applied to contexts such as Iraq and Bosnia—and if so, which.[72]

The Powell Doctrine has also seen inconsistent application in the administration of George W. Bush. In particular, the requirement for clear, finite military objectives and the assurance of a public consensus that will be sustained for the duration of the mission have not been met in Iraq. What were America's military objectives? Because the U.S. government used different arguments to justify the war to various constituencies, foreign and domestic, it came up with numerous answers to that question: it was a continuation of the fight against terrorism; it was intended to disarm Iraq; the objective was regime change; it was the beginning of democracy in Iraq and a precursor to stability throughout the Middle East. None of these outcomes would be settled, though, at the moment that the U.S. flag, the UN flag, or any other standard was raised above Saddam Hussein's numerous presidential palaces. Having removed the previous government from power and begun to install another one more agreeable to them, the United States and its allies now constituted an occupying force that would probably be required to protect the new administration and shore up Iraq's institutions for some time. This brings us to the question of public consensus: is there, and was there ever, a reasonable assurance that the U.S. public would back anything other than a short war and a quick withdrawal of American forces? These questions not only engage with the rights and wrongs of the Iraq War but also demonstrate the extent to which Secretary of Defense Donald Rumsfeld, President George W. Bush, and others in and around the current administration have been content to abandon the doctrine that bears the name of their colleague, Secretary of State Powell.

The Revival of the Munich Analogy

The Democratic Party diplomat Richard Holbrooke observed in 1987 that, whereas the central "historical myths" with which he grew up were Munich and Pearl Harbor, in the 1980s, the "enduring myths" were Vietnam and the Iranian Revolution and its aftermath. Holbrooke reported that the Vietnam precedent always weighed on the decisions of policymakers when they contemplated committing U.S. military forces. "Is every crisis Vietnam all over again? Of course not. Is there a Vietnam angle to most of our crises? To the extent that any policy involves the possible use of American forces, Vietnam always comes up."[73] But the displacement of the enduring myth of Munich by Vietnam was temporary; or perhaps one should say *intermittent*.

After the 1991 Gulf War, George Bush père famously declared "by God, we've kicked the Vietnam syndrome" and that "we've buried the specter of Vietnam in the desert sands." This statement may have been premature, because the Vietnam War continued to haunt American policy debates in the 1990s about intervention in Haiti, Colombia, Bosnia, and Kosovo. Along with the wish to bury Vietnam we have also witnessed the exhumation of the familiar skeleton, the bones of Munich. Indeed, the corpse was never completely buried. Throughout his presidency, Ronald Reagan's speeches referred frequently to the lessons of Munich in 1938, using them as proof that it was right to build America's arsenal and supply "freedom fighters" around the world, from Afghanistan to Nicaragua.[74] He accentuated the differences between Western freedom and Eastern totalitarianism, encouraging citizens of the free world to resist communism. Reagan had a soul-mate in Clinton's secretary of state, Madeleine Albright. In describing her foreign policy premises, she explained, "My mind-set is Munich," whereas "[m]ost of my generation's is Vietnam."[75]

Today, renewed use breathes new life into the Munich analogy. Members of the right-wing chorus that bolstered the Bush administration's decision to go to war in Iraq, such as Richard Perle, referred to Chamberlain's actions at Munich as the example to be avoided, and intimated that those who opposed an invasion were appeasers who, like Chamberlain, would be judged harshly by history.[76] George W. Bush implicitly referred to Munich in justifying the invasion of Iraq and the novel doctrine of "preemptive war":

> In the twentieth century some chose to appease murderous dictators whose threats were allowed to grow into genocide and global war. In this century when evil men plot chemical, biological and nuclear terror, a policy of appeasement could bring destruction of a kind never before seen on this earth . . . The security of the world requires disarming Saddam Hussein now.[77]

However, George W. Bush's presidency has by no means secured the triumph of the Munich analogy. As the U.S. occupation dragged on and a steady stream of U.S. corpses returned home in body bags, the Vietnam analogy gained new currency. Back in 1991, while politicians and pundits had voiced fears of "another Vietnam" in the prelude to "Desert Storm," these anxieties subsided after the swift withdrawal of U.S. troops following the U.S.-led coalition's victory. The pattern was reversed in the 2003 Gulf War, when the chorus of fears of "another Vietnam" *grew* after the toppling of Saddam Hussein amid the prospect of a long, perhaps costly, military occupation.

The "Lessons" of the Past

It is not always clear how politicians and their advisers use historical examples in the context of decisionmaking: do they use them *heuristically* as analytical exercises to assist them to make decisions; do they use them *didactically* as rhetorical devices to explain their decisions and persuade others; or do they use them *cosmetically* to dignify their decisions after the fact when they write their memoirs, giving their actions a learned appearance by showing how they were informed by historical knowledge?[78] When, for example, historians write that Harry Truman relied on the lessons of the 1930s in making his Korea decisions, they trust his memoirs, published some years afterward, and interviews conducted later still.[79] What is also striking is that although some historians use the Korea decision as an instance of the "lessons of Munich," Truman does not, in the memoirs, refer to Munich, but rather refers more broadly to aggression by dictators and refers specifically to Abyssinia (Ethiopia), Austria, and Manchuria. This opens up two possibilities, one of them not mentioned so far: Truman might have been retrospectively encapsulating a larger thought process in his recollection of a plane flight from Independence to Washington, when he supposedly remembered the precedents of the 1930s and resolved to fight in Korea; or, in their attempt to give the past a pattern, historians may have seized on this moment to make sense of Truman's decisionmaking process. The significance of the Munich precedent may be as much a historiographical artifact as a historical fact. This leaves a question mark over how large a part the lessons of the 1930s played in Truman's decision.

In contrast, records of the discussions of 1964 and 1965 demonstrate that Johnson and his advisers used the Munich analogy when they were discussing Vietnam at the time among themselves.[80] The Munich analogy was not, therefore, solely for public consumption nor was it a later gloss they added in their memoirs. This suggests that the analogy was a component of the decisionmaking process. It would be a mistake, though, to separate too rigidly the heuristic (decisionmaking) and didactic (decision-explaining) function of the analogy—both presume the "fit" between a historical precedent and a contemporary situation, and one must first persuade oneself of the usefulness of an analogy before using it to persuade others. The use of the Munich analogy in the councils of government suggests, though, that it was more than a presentational device employed to explain a decision to the public at the time or to rationalize it after the fact.

By the time we get to the 1980s, the role of the Vietnam precedent becomes enormously complicated. Arguments about the right course of action in Central America intermingled with arguments about what had gone wrong in Vietnam, and it was hardly possible to exchange words about the "lessons" of Vietnam without implicitly debating Central America policy. The Vietnam precedent was not, therefore, a heuristic device that one could explore in the spirit of discovery in order to draw lessons about Central America. The debates about Vietnam and Central America were wrapped together in a single package. Policymakers' views about both must have emerged from and conformed to the policymakers' overall worldview. If a policymaker saw international politics in terms of communist aggression, this worldview was likely to produce congruent interpretations of Central America in the 1980s and Southeast Asia in the 1960s; if a policymaker saw international politics as an arena where the United States tended to overreach, this would produce another coherent view of the U.S. role in the two regions. The history of the Vietnam War might have helped both policymakers to arrive at their worldviews; but it was also possible that subsequent events might reinforce—or undermine—their interpretations of the past. However the "lessons" of Vietnam might have served in all this, they do not conform cleanly to the heuristic or didactic types. The Vietnam War instead became part of an expanded arena in which ideological conflict between advocates of different policies took place—a theater of combat comprising both past and present and encompassing the world.

Ernest May has argued "framers of foreign policy are often influenced by beliefs about what history teaches or portends." Unfortunately, May remarks, "policy-makers ordinarily use history badly. When resorting to an analogy, they tend to seize upon the first that comes to mind. They do not search more widely. Nor do they pause to analyze the case, test its fitness, or even ask in what ways it might be misleading. Seeing a trend running toward the present, they tend to assume that it will continue into the future, not stopping to consider what produced it or why a linear projection might be mistaken."[81] While this may be true, it does not explain why one historical event rather than another is "the first that comes to mind." Why do some experiences bear a heavier weight than others and press upon the mind more deeply or insistently?

Just as traumatic events impress themselves powerfully on individual psyches, the events that inscribe themselves most deeply into the minds of policymakers—and the societies they represent—are usually catastrophes.[82] After World War II, Pearl Harbor reminded Americans of the danger of sudden attack, and made "readiness" a watchword not just for the military but for a society that had seen the terrible destruction that air power could wreak. This lesson was an adjunct to the strategic lessons that American policymakers drew from Chamberlain's meeting with Hitler at Munich in 1938. Munich was the ruling analogy for the generation that made policy from 1945 to 1970 not because of its intrinsic applicability to the cold war context but simply because the warning it seemed to sum up carried all the devastating weight of the recent world war.

Was the Munich precedent an appropriate analogy for Southeast Asia?[83] To what subsequent situations can the "lessons" of Vietnam be applied? These questions require us to think about the basis on which any historical experience can be a useful analogy for another. In 1967, James C. Thomson anticipated the eventual lesson learning that would follow the Vietnam War. Suggesting that the Vietnam War was historically unique, he mischievously described its central lesson as follows: "Never again to take on the job of trying to defeat a nationalist anticolonial movement under indigenous communist control in former French Indochina."[84] In other words, the Vietnam War is a sui generis experience with no relevance for any other. Historians can hardly accept such rigorous agnosticism about the application of the "lessons" of the past. After all, we never cease to decry the lack of historical knowledge and insight of politicians and the general public and there is not much point in learning about the past if it remains so radically discontinuous from the present that it offers no guidance or warning.

Analogies depend on the discovery of identity across differences—they are extended metaphors. We can agree that no two historical events or situations are identical. The formulation of an analogy requires us to identify the features of one historical experience that can be applied to another, and the discovery of their "fit" is always a matter of selection and is open to debate. The selection of one past event over another as the model on which the present can be conceived will highlight one set of features among a limitless universe of possibilities. Just as the selection of a past historical experience affects how one sees the present, the specific features of the present in which one is interested will suggest one past model over another. The modeling of the plausible "fit" between past and present is thus a two-way process.

If, for example, successive exercises in "nation building" come off successfully, we are likely to forget the "lesson" of Vietnam that said nation-building must be avoided. Whenever nation-building founders, as when the mission in Somalia ended ignominiously, that "lesson" of the Vietnam War will be reinforced. As one Democratic party Senator said after an American force suffered casualties in Somalia, "It's Vietnam all over again." Henry Hyde, a Republican member of the House Foreign Affairs Committee, chimed in that the deployment of a small number of U.S. troops was "compounding the Johnsonian error in Vietnam of incrementally deploying forces."[85] What transpires in a changing present continually refreshes and renews the past—both by providing an ever-expanding past from which to draw new lessons and on which to construct new analogies, and by reinscribing old lessons and analogies that seem to remain relevant. The debacles in Lebanon and, a decade later, in Somalia, retaught some old lessons about the risks of using force and sending troops in small numbers to intervene in complex local conflicts. But they also provided a fresh idiom in which to express these familiar warnings, allowing pundits to refer to the "lessons" of Lebanon and Somalia.[86]

Before the official beginning of the 2003 invasion of Iraq, I said the following in a teach-in about the prospective war:

> If we accept the applicability of the Munich precedent to Iraq, it obliges our leaders to say that they will be steadfast and ignore the opinions of their populace; the biggest danger, though, is that they might actually mean this. The incalculability of the risks on both sides of the case—the danger that dissent in Britain and the US might cause some miscalculation on Iraq's part; the danger that our leaders may drag us willy nilly into a war that few of us want—is one of the things that is making this political moment so fraught. If our leaders really, really intend to go to war without the full backing of their populations, the Munich analogy may not survive the first shots in a war that is bound to lead to civilian casualties and to the loss of life in all the armed forces involved. Even if victory comes swiftly to the allies, and there is no certainty that it will, we must be prepared to contemplate a reconstruction and occupation of a country whose population may not greet the foreigners as liberators but may view them as invaders. If this occurs, the specter of the Vietnam War may return to haunt us all over again.[87]

When the statue of Saddam Hussein toppled in Baghdad, this began to seem an absurd formulation, and I was relieved that I had hedged against various possibilities. As time passed, though, the hedges seemed to become less necessary and my fears of being trumped by history receded. How valid the comparisons will appear—and what the invasion and occupation of Iraq will teach—depends, as all exercises in "lesson learning," on the eventualities of an unpredictable future.

Notes

1. Maureen Dowd, "What Tips for Rumsfeld from Julius Caesar?," *International Herald Tribune*, March 6, 2003, 9.
2. Michael T. Kaufman, "Pentagon Film Group Watches Algiers While Thinking Iraq," *International Herald Tribune*, September 8, 2003, 2; for a report of the Nixon White

House's viewing of the film, see Thomas W. Pauken, *The Thirty Years War: The Politics of the Sixties Generation* (Ottawa, IL: Jameson Books, 1995), 105–106.

3. Ernest R. May, *The "Lessons" of the Past: The Use and Misuse of History in American Foreign Policy* (New York: Oxford University Press, 1973), 81–82; Richard A. Neustadt and Ernest R. May, *Thinking in Time: The Uses of History for Decision Makers* (New York: Free Press, 1986), 41; Loren Baritz, *Backfire: A History of How American Culture Led Us into Vietnam and Made Us Fight the War We Did* (Baltimore and London: The Johns Hopkins University Press, 1998), 72–73; Joseph Smith, *The Cold War, 1945–1991* (2nd edn.; Oxford: Blackwell, 1998), 59.

4. Yuen Foong Khong, *Analogies at War: Korea, Munich, Dien Bien Phu and the Vietnam Decisions of 1965* (Princeton: Princeton University Press, 1992), 77.

5. Walter L. Hixson, *George F. Kennan: Cold War Iconoclast* (New York: Columbia University Press, 1989), 153.

6. Khong, *Analogies at War*, 178.

7. May, *The "Lessons" of the Past*, 96–97.

8. Neustadt and May, *Thinking in Time*, 87.

9. Brian VanDeMark, *Into the Quagmire: Lyndon Johnson and the Escalation of the Vietnam War* (New York and Oxford: Oxford University Press, 1995), 191, 215; H.R. McMaster, *Dereliction of Duty: Lyndon Johnson, Robert McNamara, and the Joint Chiefs of Staff, and the Lies that Led to Vietnam* (New York: HarperCollins, 1997), 146; Khong, *Analogies at War*, 3.

10. May, *The "Lessons" of the Past*, 114.

11. "We Will Stand in Vietnam," *Department of State Bulletin*, August 16, 1965, 262, cited in Khong, *Analogies at War*, 179.

12. *Why Vietnam* (Washington, D.C.: U.S. Directorate for Armed Forces Information and Education, 1965); copy supplied to the author by the Lyndon Baines Johnson Presidential Library, Austin, Texas.

13. Gerald R. Ford, "Remarks at Veterans Day Ceremonies at Arlington National Cemetery," October 28, 1974, *Public Papers of the Presidents: Gerald Ford, 1974* (Washington, D.C.: U.S. Government Printing Office, 1975), 477.

14. Walter Isaacson, *Kissinger: A Biography* (London: Faber and Faber, 1993), 611.

15. Randall Bennett Woods, *J. William Fulbright, Vietnam, and the Search for a Cold War Foreign Policy* (Cambridge: Cambridge University Press, 1998), 147; Richard A. Melanson, *Writing History and Making Policy: The Cold War, Vietnam, and Revisionism* (Lanham, New York, and London: University Press of America, 1983), 212.

16. Paul Seabury, "The Moral and Strategic Lesson of Vietnam," in John Norton Moore, ed., *The Vietnam Debate: A Fresh Look at the Arguments* (Lanham, MD: University Press of America, 1992), 13.

17. Gloria Emerson, *Winners and Losers: Battles, Retreats, Gains, Losses and Ruins from the Vietnam War* (New York and London: Harcourt Brace Jovanovich, 1976), 36.

18. Gerald Ford, "The President's News Conference, May 6, 1975," *Public Papers of the Presidents Gerald Ford, 1975, Book I—January 1 to July 17, 1975* (Washington, D.C.: U.S. Government Printing Office, 1977), 645.

19. Melvin Small, *The Presidency of Richard Nixon* (Lawrence, KS: University Press of Kansas, 1999), 63.

20. Isaacson, *Kissinger*, 677; Thomas G. Paterson, *Meeting the Communist Threat: Truman to Reagan* (New York: Oxford University Press, 1988), 245.

21. Small, *Presidency of Richard Nixon*, 63.

22. Editorial, *The Washington Post*, January 6, 1975; cf. *The Roanoke Times*, January 8, 1975; *The Burlington [Vermont] Free Press*, January 4, 1975. *Editorials on File*, 21.

23. Press conferences of April 3, 1975; April 16, 1975; and April 21, 1975: *Public Papers of the Presidents: Gerald Ford, 1975, Book I,* 420–421, 505, 544; William Turley, *The Second Indochina War: A Short Political and Military History, 1954–1975* (New York and Scarborough, Ontario: New American Library, 1987), 183; Marilyn B. Young, *The Vietnam Wars, 1945–1990* (New York: HarperCollins, 1991), 291.

24. Stanley Karnow, *Vietnam: A History* (New York: Viking, 1983), 667.

25. Arnold R. Isaacs, *Without Honor: Defeat in Vietnam and Cambodia* (London and Baltimore: Johns Hopkins University Press, 1983), 500.

26. Jeffrey Kimball, *Nixon's Vietnam War* (Lawrence: University Press of Kansas, 1998), 371.

27. Michael Lind, *Vietnam: The Necessary War: A Reinterpretation of America's Most Disastrous Military Conflict* (New York: Touchstone Books, 1999), 81.

28. See, e.g., Karnow, *Vietnam,* 17.

29. Melanson, *Writing History and Making Policy,* 199.

30. Norman Podhoretz, *Why We Were in Vietnam* (New York: Simon and Schuster, 1984), 200.

31. Khong, *Analogies at War,* 258.

32. Howell Raines, "Reagan Calls Arms Race Essential to Avoid a 'Surrender' or 'Defeat,' " *New York Times,* August 19, 1980, D17.

33. Lou Cannon, "Reagan: 'Peace through Strength,' " *Washington Post,* August 19, 1980, A4.

34. Isaacs, *Without Honor,* 488.

35. *Public Papers of the Presidents, Ronald Reagan, 1981—January 20 to December 31, 1981* (Washington, D.C.: U.S. Government Printing Office, 1982), 155; Walter Capps, *The Unfinished War: Vietnam and the American Conscience* (rev. edn.; Boston: Beacon Press, 1990), 146; Ronald Reagan, Press Conference, April 18, 1985, *Public Papers of the Presidents, January 1 to June 28, 1985* (Washington, D.C.: U.S. Government Printing Office, 1986), 454.

36. Jan Scruggs and Joel Swerdlow, *To Heal a Nation* (New York: Harper and Row, 1985), 152.

37. Footage of the speech is shown in the documentary, *Frontline: Vietnam Memorial* (producers, Steve York and Foster Wiley, produced by WGBH for the documentary consortium, 1983; broadcast on the PBS affiliate, WNET, New Jersey, May 28, 1991).

38. Helga Haftendorn, "Toward a Reconstruction of American Strength: A New Era in the Claim to Global Leadership?" in Helga Haftendorn and Jakob Schissler, eds., *The Reagan Administration: A Reconstruction of American Strength?* (Berlin, New York: de Gruyter, 1988), 3.

39. Quoted in Melanson, *Writing History and Making Policy,* 203.

40. Lou Cannon, *President Reagan: The Role of a Lifetime* (New York: Public Affairs, 2000), 390.

41. Jane Mayer and Doyle McManus, *Landslide: The Unmaking of the President, 1984–1988* (Glasgow: Fontana/Collins, 1989), 31.

42. *The Denver Post,* August 26, 1980; *Editorials on File,* 996.

43. *Arizona Republic,* August 21, 1980; *Editorials on File,* 993.

44. *The Miami Herald,* August 20, 1980; *Editorials on File,* 996.

45. *The Chicago Sun-Times,* August 21, 1980; *Editorials on File,* 992.

46. Antonio Gramsci, "The Intellectuals," in *Selections from the Prison Writings* (New York: International Publishers, 1971); Antonio Gramsci, "Hegemony, Relations of Force, Historical Bloc," in *A Gramsci Reader* (New York: Lawrence and Wishart, 1988), 189–245.

47. William M. LeoGrande, *Our Own Backyard* (Chapel Hill and London: University of North Carolina Press, 1998), 97. For recurrent instances where the specter or lessons of Vietnam impinged on debate about Central America policy, see 75–103.

48. LeoGrande, *Our Own Backyard,* 140; Kathryn Roth and Richard Sobel, "Chronology of Events and Public Opinion," in Richard Sobel, ed., *Public Opinion in U.S.*

Foreign Policy: The Controversy over Contra Aid (Lanham, MD: Rowman and Littlefield, 1993), 22.

49. Gallup Organization poll for *Newsweek* magazine, February 17–18, 1982. Unless otherwise cited, all poll results were obtained from the Roper Center for Public Opinion Research, University of Connecticut, Storrs.

50. Roth and Sobel, "Chronology of Events and Public Opinion," 23.

51. James M. Scott, *Deciding to Intervene: The Reagan Doctrine and American Foreign Policy* (Durham and London: Duke University Press, 1996), 163.

52. Scott, *Deciding to Intervene*, 163.

53. LeoGrande, *Our Own Backyard*, 303.

54. Ibid., 189.

55. Gallup polls, March 13–16, 1981; June 24–27, 1983; July 29–August 2, 1983; September 9–12, 1983; May 18–21, 1984; Gallup poll for *Newsweek* magazine, February 17–18, 1982; Harris polls, March 12–16, 1982; April 29–May 1, 1983; August 18–22, 1983.

56. Scott, *Deciding to Intervene*, 165.

57. Christian Smith, *Resisting Reagan: The U.S. Central America Peace Movement* (Chicago and London: University of Chicago Press, 1996), 94; Michael Klare, *Beyond the "Vietnam Syndrome": U.S. Interventionism in the 1980s* (Washington, D.C.: Institute for Policy Studies, 1981), 95.

58. LeoGrande, *Our Own Backyard*, 244, 340.

59. Mark Hertsgaard, *On Bended Knee: The Press and the Reagan Presidency* (New York: Farrar, Straus Giroux, 1988), 271.

60. Cannon, *President Reagan: The Role of a Lifetime*, 310, 332.

61. Gallup polls for the Chicago Council on Foreign Relations: November 17–26, 1978; October 29–November 6, 1982; and October 30–November 12, 1986.

62. CBS News/New York Times polls, February 23–27, 1985, and May 29–June 2, 1985.

63. Robert J. McMahon, *The Limits of Empire: The United States and Southeast Asia Since World War II* (New York: Columbia University Press, 1999), 183.

64. Richard Nixon, *No More Vietnams* (New York: Arbor House, 1985), 47, 237.

65. Kim Willenson, *The Bad War: An Oral History of the Vietnam War* (New York and Scarborough, Ontario: New American Library, 1987), 391–392.

66. Ibid., 406.

67. Colin Powell, *A Soldier's Way: An Autobiography* (London: Hutchinson, 1995), 434.

68. Caspar Weinberger, "The Uses of Military Power" [based on a speech first delivered to the National Press Club in November 1984], *Defense Issues* 2: 44 (January 1985), published by the American Forces Information Service, Office of the Assistant Secretary of Defense (Public Affairs).

69. Boot questions whether vital interests were at stake in the Barbary Pirates War, the Boxer rebellion of 1900, and the invasion of Veracruz in 1914. He points out that there was no exit strategy in the occupations of Haiti from 1915 to 1933, of Nicaragua from 1910 to 1933, and of the Philippines from 1898 to 1946. U.S. troops acted as colonial administrators in the Philippines, Haiti, Cuba, Nicaragua, Veracruz, post–World War II Germany, Japan, and the post–Civil War South. Max Boot, "Everything You Think You Know About the American Way of Fighting War Is Wrong," *History News Network*, October 7, 2002. http://hnn.us/articles/1022.html consulted on October 11, 2002.

70. Michael T. Klare and Peter Kornbluh, "The New Interventionism: Low-Intensity Warfare in the 1980s and Beyond," in Klare and Kornbluh, eds., *Low-Intensity Warfare: Counterinsurgency, Proinsurgency, and Antiterrorism in the Eighties* (New York: Pantheon, 1988), 3–9.

71. Scott, *Deciding to Intervene*, 22.
72. Mary McGrory, "Vietnam or Munich?" *The Washington Post*, June 6, 1995, A2; Thomas L. Friedman, "American Power and an Envious, Angry World," *International Herald Tribune*, March 2, 1998, 8.
73. Quoted in Willenson, *The Bad War*, 397.
74. Interview with Walter Cronkite, March 3, 1981, *Public Papers of the Presidents: Ronald Reagan, January 20 to December 31, 1981* (Washington, D.C.: U.S. Government Printing Office, 1982), 194–195; Interview with Foreign Journalists, May 26, 1983, *Public Papers of the Presidents: Ronald Reagan, 1983, Book I—January 1 to July 1, 1983* (Washington, D.C.: U.S. Government Printing Office, 1984), 774; "Radio Address to the Nation on Defense Spending," *Public Papers of the Presidents: Ronald Reagan, 1983, Book I—January 1 to July 1, 1983*, 257; Meeting with student leaders, *Public Papers of the Presidents: Ronald Reagan, 1983, Book I—January 1 to July 1, 1983*, 948–949.
75. E.J. Dionne, "Don't Applaud Albright Just Because She's a She," *International Herald Tribune*, December 11, 1996, 11; Nancy Gibbs, "The Many Lives of Madeleine," *Time* 149: 7 (February 17, 1997), 36, 41.
76. Steven C. Day, "Iraq and the Ghosts of Munich: A Popular Lesson is Misapplied," http://www.poppolitics.com/articles/2003-01-28-ghostsofmunich.shtml, consulted March 10, 2003; David Horowitz, "Left Scrambles to Betray America," http://www.newsmax.com/archives/articles/2002/12/31/174634.shtml, consulted March 10, 2003.
77. "Text: Bush's Speech on Iraq," *New York Times*, March 18, 2003; found at http://www.nytimes.com/2003/03/18/politics/18BTEX.html?pagewanted=3
78. Khong, *Analogies at War*, 14–16.
79. May, *The "Lessons" of the Past*, 81–82.
80. McMaster, *Dereliction of Duty*, 248; Khong, *Analogies at War*, 178.
81. May, *The "Lessons" of the Past*, ix, xi; Cf. Neustadt and May, *Thinking in Time*, 89.
82. Edward Linenthal, *Sacred Ground: Americans and Their Battlefields* (Urbana and Chicago: University of Illinois Press, 1993), 63, 178.
83. The question was posed sharply by Earl C. Ravenal, *Never Again: Learning from America's Foreign Policy Failures* (Philadelphia: Temple University Press, 1978), 33–35.
84. Richard M. Pfeffer, ed., *No More Vietnams? The War and the Future of American Foreign Policy* (New York, Evanston, and London: Harper and Row, 1968), 258.
85. Clifford Krauss, "White House Tries to Calm Congress," *The New York Times*, October 6, 1993, 16.
86. Report by Jack Smith, referring to the "lessons" of Somalia and Lebanon, but not Vietnam, *This Week with David Brinkley*, ABC television network, April 17, 1994.
87. Teach-in at Lancaster University's Furness College, March 10, 2003.

Suggestions for Further Reading

Appy, Christian G., *Patriots: The Vietnam War Remembered from All Sides* (New York: Viking, 2003).

Franklin, H. Bruce, *M.I.A. or Mythmaking in America* (expanded and updated edition; New Brunswick, NJ: Rutgers University Press, 1993).

Hendrickson, Paul, *The Living and the Dead: Robert McNamara and Five Lives of a Lost War* (New York: Knopf, 1996).

Isaacs, Arnold R., *Vietnam Shadows: The War, Its Ghosts, and Its Legacy* (Baltimore and London: Johns Hopkins University Press, 1997).

Lind, Michael, *Vietnam: The Necessary War: A Reinterpretation of America's Most Disastrous Military Conf* (New York: Touchstone, 1999).

Neu, Charles E., ed., *After Vietnam: Legacies of a Lost War* (Baltimore and London: Johns Hopkins University Press, 2000).

Nicosia, Gerald, *Home to War: A History of the Vietnam Veterans' Movement* (New York: Three Rivers Press, 2001).

Timberg, Robert, *The Nightingale's Song* (New York: Simon and Schuster, 1995).

Young, Marilyn B., *The Vietnam Wars, 1945–1990* (New York: HarperCollins, 1991).

Acknowledgments

The editors would like to express their appreciation to Tony Judt and Jair Kessler of New York University's Remarque Center for their generosity in organizing a conference in October 2003, bringing together participants in this volume with colleagues interested in the politics of the past. In that setting, Ian Buruma, Katherine Fleming, Sunil Khilnani, Charles King, Mark Mazower, Andrew Shennan, and Marilyn Young provided thoughtful commentary and criticism at a crucial juncture in the development of the book. David Pervin at Palgrave Macmillan developed the original idea and was supportive and gracious throughout the process, granting editors and authors complete independence. The editors would like to thank all the contributors for their insights and expertise, and for their willingness to submit to the editorial demands of two taskmasters. In addition, Max Paul Friedman thanks Joan Casanovas, Michael Creswell, Martin Friedman, Christopher Griffin, Dirk Moses, Steve Stern, and Katharina Vester, and Padraic Kenney thanks Izabela Ziólkowska-Kenney, and the students in his International Affairs seminar on the post–cold war world at the University of Colorado in Fall 2001.

List of Contributors

Subho Basu is assistant professor in South Asian History at Illinois State University.

Andrew H. Beattie is associate lecturer in German Studies at the Institute for International Studies of the University of Technology, Sydney.

Suranjan Das is pro vice-chancellor for Academic Affairs and professor of History at the University of Calcutta and honorary director of the Netaji Institute for Asian Studies in Kolkata, India.

Alexis Dudden is Sue and Eugene Mercy associate professor of History at Connecticut College.

Toyin Falola is Frances Higginbothom Nalle Centennial professor in History at the University of Texas at Austin.

Max Paul Friedman is assistant professor of History at Florida State University.

Patrick Hagopian is lecturer in American Studies at Lancaster University.

Katherine Hite is associate professor of Political Science at Vassar College.

Carsten Jacob Humlebæk is assistant professor of Spanish culture in the Department of French, Italian, Russian, Spanish, and German at Copenhagen Business School.

Padraic Kenney is professor of History at the University of Colorado at Boulder.

Ilan Pappe is senior lecturer in Political Science at Haifa University.

Ronald Grigor Suny is professor of Political Science and History at the University of Chicago.

Index

Entries in **boldface** refer to chapters devoted to that particular country.